Senescence

Dominant or Recessive in Somatic Cell Crosses?

CELLULAR SENESCENCE AND SOMATIC CELL GENETICS

Series editors: **Warren W. Nichols**
Institute for Medical Research
Camden, New Jersey

Donald G. Murphy
National Institute on Aging
Bethesda, Maryland

Volume 1 REGULATION OF CELL PROLIFERATION AND DIFFERENTIATION
Edited by Warren W. Nichols and Donald G. Murphy

Volume 2 SENESCENCE: DOMINANT OR RECESSIVE IN SOMATIC CELL CROSSES?
Edited by Warren W. Nichols and Donald G. Murphy

Volume 3 MYCOPLASMA INFECTION OF CELL CULTURES
Edited by Gerard J. McGarrity, Donald G. Murphy, and Warren W. Nichols

Senescence

Dominant or Recessive in Somatic Cell Crosses?

Edited by

Warren W. Nichols

Institute for Medical Research
Camden, New Jersey

and

Donald G. Murphy

National Institute on Aging
Bethesda, Maryland

PLENUM PRESS · NEW YORK AND LONDON

Library of Congress Cataloging in Publication Data

Main entry under title:

Senescence: dominant or recessive in somatic cell crosses?
(Cellular senescence and somatic cell genetics; 2)
"Proceedings of the Institute for Medical Research symposium."
Includes index.
1. Cells—Aging—Congresses. 2. Cytogenetics—Congresses. 3. Somatic hybrids—Congresses. I. Nichols, Warren W. II. Murphy, Donald G. III. Institute for Medical Research, Camden, N. J. IV. Series. [DNLM: 1. Cytogenetics—Congresses. 2. Cells, Cultured—Congresses. 3. Aging—Congresses. W1 CE1295 v. 2 1976/WT104 S475 1976]
QH608.S46 574.8'761 77-25004
ISBN 978-1-4684-2510-9 ISBN 978-1-4684-2508-6 (eBook)
DOI 10.1007/978-1-4684-2508-6

Proceedings of the Institute for Medical Research Symposium held in May 1976

Softcover reprint of the hardcover 1st edition 1977
A Division of Plenum Publishing Corporation
227 West 17th Street, New York, N.Y. 10011

Organizing Panel

Donald G. Murphy, Health Scientist Administrator
National Institute on Aging, National Institutes of Health, Bethesda, MD

Warren W. Nichols, Head, Dept. of Cytogenetics
Institute for Medical Research, Camden, NJ

Participants

Vincent J. Cristofalo, The Wistar Institute of Anatomy and Biology, Philadelphia, PA

Carlo M. Croce, The Wistar Institute of Anatomy and Biology, Philadelphia, PA

Richard L. Davidson, Harvard Medical School, Boston, MA

Robert DeMars, University of Wisconsin, Madison, WI

John W. Littlefield, Johns Hopkins University School of Medicine, Baltimore, MD

Orlando J. Miller, College of Physicians & Surgeons, Columbia University, New York, NY

Audrey L. Muggleton-Harris, University of Massachusetts Medical School, Worcester, MA

Thomas H. Norwood, University of Washington, Seattle, WA

F. Marott Sinex, Boston University School of Medicine, Boston, MA

Rosalind Yanishevsky, University of Colorado, Boulder, CO

Preface

This monograph, Senescence: Dominant or Recessive In Somatic Cell Crosses? represents the second annual workshop to promote theory and concept development in aging research. These workshops are part of a resource to bank cultured cells of special interest to aging research that was established at the Institute for Medical Research in Camden, New Jersey, by the National Institute on Aging in 1974. The underlying theme of the workshops is the use of cultured cells in a variety of somatic cell genetic systems designed to define mechanisms of *in vitro* cellular scenescence and the possible insights that this may provide to the problems of *in vivo* aging. The concept also includes bringing together workers from a variety of disciplines to stimulate new and innovative thoughts and work in the area. The current workshop focuses on the relative role of nucleus and cytoplasm on determining the *in vitro* lifespan of human diploid cells as well as the relative influence of old and young cells when combined within a single cell structure. The techniques and procedures discussed should make significant contributions to understanding *in vitro* senescence and may lead to the mapping of an area or areas of the genome linked to senescence as is being accomplished with viral transformation of normal cells.

Warren W. Nichols
Donald G. Murphy

Contents

THEORETICAL MECHANISMS OF IN VITRO SENESCENCE

F. Marott Sinex

Boston University School of Medicine

80 East Concord St., Boston, Mass. 02118

In this chapter we will discuss how somatic mutations might be involved in the senescence of cells in culture and in the intact organism. We will also discuss the possible role that stable phenotypic variations may play in senescence.

Strictly speaking somatic mutations refer to changes in DNA which alter the genetic information available to a cell and which are capable of transmission to the cells progeny. In culture it is sometimes difficult to tell whether DNA is truly altered or whether a relatively stable change in regulation or differentiation has occurred.

Without knowing what is altered it is difficult to really understand mechanisms, or to conceptualize means of intervention. If aging is indeed due to mutation why does the mutation occur and how can it be prevented? Is the mutation intrinsic to cell replication or are base pairs actually altered chemically, even in the resting state? On the other hand, might not stable phenotypic variants arise through altered cellular regulation or differentiation in a cell culture which would be propagated?

What can be demonstrated in tissue culture is the following: Certain cell types, but thus far by no means all, can be propagated. Different diploid cell types will undergo a variable number of divisions characteristic of the tissue or origin. After a number of population doublings the cells usually die or transform. Human fibroblasts may be passaged a relatively large number of times before dying out, and show little tendency to transform, differing from rodent fibroblasts which transform much more readily. We can observe how long cells can proliferate during a single passage, and how many doublings they will undergo. We can estimate how many cells in the population appear to be cycling and study the variance in cycling or not cycling, and the time between divisions for individual cells and for the population as a whole. We can study what affects these divisions, such as subcultivation ratio, and variation in the concentration of calf serum, glucocorticoid, insulin and arachadonic acid in the culture media. We can contrast the behavior of cells cultured from normal individuals of different ages as well as cells from subjects with progeria, diabetes and Xeroderma pigmentosum (Goldstein & Niewiarowski, 1976). We can infect the cells with different viruses, both DNA and RNA, and study transformation and DNA injury and repair. Finally, one can hybridize young and old cells, and insert young and old nuclei into young and old cytoplasm, to learn whether old nuclei are dominant over the nuclei of young or malignant cells. Up to now, many gerontologists have perhaps emphasized the entrance of cells into DNA synthesis in early and late passage cells rather than evidence for and against mutation and epigenetic change which is now emerging.

There are two bodies of evidence relating to the importance of somatic mutation in aging, one based on the study of intact organisms, and the other of tissue cultures *per se*. Howard Curtis was the first to vigorously espouse mutation of DNA as the cause of aging. Curtis (1966) carried out a chemical partial hepatectomy by exposing animals to carbon tetrachloride. This induced scorable mitoses in the livers which enabled him to follow chromosomal aberrations. Curtis found that these increased significantly with age. The number of cells showing at least one chromosomal aberration was high, being as high as 45 percent for mice and 33 percent for guinea pigs (Curtis & Miller, 1971). The rate of appearance of the aberrations was correlated with life span, being faster in mice than in longer lived guinea pigs and dogs. Strehler (1959) repeatedly pointed out in discussions of the Curtis work that ionizing radiation was much more efficient in producing chromosomal aberrations than life shortening. Curtis's own data showed that by the time irradiated animals evidenced life shortening, most of the chromosomal aberrations induced by ionizing radiation had disappeared. However, chromosomal aberrations may be but the tip of the mutational iceberg. The significance of induced chromosomal damage may be its high correlation with gene mutations (Kihlman, 1966).

Curtis recognized the analogy between aging and the effects of chronic low level radiation. Animals irradiated earlier have shorter lives and appear older later on with respect to coats, behavior and susceptibility to disease. Since radiation injury was presumed to occur through mutation of DNA and aging resembled chronic low level radiation injury, it was natural to postulate that aging resulted from mutational injury. The problem here is that radiation injury may not be completely analogous to aging (Upton, 1957, 1977). For example, both aging and radiation injury may affect the microcirculation (Casarett, 1964).

Lymphocytes from older human subjects stimulated with phytohemagglutinin show an increasing frequency of aberrations (Jacobs, et al., 1961, 1963). This is particularly true with patients with organic brain syndrome (Jarvik et al., 1974). The most common aberration is the loss of an X chromosome in females, or some other form of hypodiploidy.

Interest in somatic mutation as a cause of aging was stimulated when Hart and Setlow (1974) made a rather striking discovery, namely that the efficiency of the repair enzymes of DNA seemed to be correlated with species longevity. The original technique of Hart and Setlow involved autoradiographic determination of unscheduled DNA synthesis using ^{3}H-thymidine. It is tempting to consider repair enzymes important in determining longevity and by inference the injury to DNA which required repair. Long-lived species appeared considerably more efficient in excising induced dimers than the short-lived ones.

Burnet (1975) took another approach in his discussion on intrinsic mutagenesis. He argued that somatic mutation is brought about by error in DNA synthesis associated with cell replication, the error level in a particular tissue being determined by the properties of its DNA polymerase.

Before discussing evidence for somatic mutation from experiments in culture let us consider some of the evidence against somatic mutation as a cause of aging in the intact animal. Alexander (1960) attempted to demonstrate geromimetic effects with bifunctional alkylating agents and was discouraged with his results. He ran into considerable toxicity, and while he did obtain some life shortening he felt that this could best be attributed to cellular loss. He went further than this, however, in his interpretation and suggested that his experiments disproved the importance of somatic mutation in aging, which was hardly the case. However, this interpretation worked its way into the reviews of the period and gained some acceptance.

Aging is not correlated with ploidy. Polyploidy gives only partial protection against either radiation or aging. In the most straightforward approach, if radiation injury or aging were due to recessive mutations, then haploid animals should be short-lived and polyploid organisms exceptionally long-lived.

It was Lamb and Maynard Smith (1965) working with *Drosophila*, and Clark and Rubin (1961) working with a wasp, *Habrobracon sp.*, who raised the issue of whether aging hits are dominant or recessive. Leo Szilard (1959) had suggested that aging injury was recessive. In Szilard's view aging hits were seen as affecting a locus with a previously existing mutational burden at the homologous locus. If we assume that aging mutations are random then the probability of obtaining a somatic mutation that is a match for pre-existing recessive mutational burden is small. For example, if there are at least 10,000 loci in a haploid set, the probability of a gene match is 1/10,000. If aging mutations were recessive then there should be an enormous variation in longevity with ploidy. Since there is none in these insect systems the situation seems to be more complex. The post-mitotic state of the adult somatic cells of insects must be considered since cell division is important to both the production and removal of mutations as well as the possibility that aging death in insects may have rather specialized mechanisms.

Other possible explanations for the lack of correlation between ploidy and life span are: (1) That the majority of aging mutations are expressed as dominants rather than recessives; (2) that what are actually involved are stable phenotypic variants, and that by whatever mechanisms these arise and are maintained, ploidy has little effect.

Now let us consider specifically what happens in cell cultures. We have always felt that somatic mutations must occur in the tissue of long-lived animals. The question seemed to be how significant is this for aging. We have reviewed the frequency of somatic mutation in tissue culture to see if this was sufficient to account for the changes observed in aging tissue (Sinex, 1974). The mutation rate in culture appears to be around 4.5×10^{-6}/gene/generation or per two or three days. If the number of doublings represented by various cells in the body were known one could multiply the number of mutations per generation by the number of genes times the average doublings which a cell has undergone to obtain an estimate of the number of possible mutations per cell in an organism. If we very roughly take 2×10^4 for the number of genes and something like 34 generations for the average cell, then $4.5 \times 10^{-6} \times 34$ yields something like three mutations in an average cell per lifetime. The number 34 is a very crude approximation. No one ever counted the number of cells in the body, averaged their doublings, or averaged their size. Nevertheless, we are dealing with exponents, and a proper number should be somewhere between 30 and 50. If on the other hand you take mutation in culture on a time basis and consider that a human's lifetime of 25,000 days is something like 10,000 times as long as the time between doublings, then $4.5 \times 10^{-6} \times 2 \times 10^4 \times 1 \times 10^4$ would give something like 900 mutations per cell per lifetime. Therefore experiments in tissue culture would suggest that aging via somatic mutations is perfectly feasible.

But what about phenotypic variation? It is recognized that it can be very difficult to distinguish between mutation and stable phenotypic variation. What can experimentation in tissue culture offer towards a solution of this problem? This brings us back to discussing a series of experiments concerning dominant and recessive genes and ploidy. As we explained in our discussion of ploidy, the expression of recessive genes in haploid, diploid, tetraploid and octaploid cells should vary exponentially. Two groups have compared apparent mutation rates in cells of varying ploidy in culture. Harris (1971) studied the expression of mutations in diploid, tetraploid and octaploid Chinese hamster cells. He studied resistance to azaguanine and to heating to 43° as an expression of mutation. He observed that tetraploid cells showed roughly twice the number of mutations as diploid cells while octaploids showed only half as many. That is, he found more mutations than would be expected for recessives in the tetraploid cells and fewer mutations than would be expected for dominant effects with increasing ploidy in the system.

To further illustrate the effects of ploidy, Mezger-Freed (1972) worked with haploid cells from embryonic frogs. Her particular line, ICR-2A, was the first haploid vertebrate cell to be established. The mutation followed was the loss of thymidine kinase activity. When thymidine kinase disappears, the cells can no longer incorporate bromodeoxyuridine (BrdU). Cells which contain BrdU may be destroyed by exposure to ultraviolet light. The surprising result was that the frequency of mutation in haploid cells was 5 percent that in diploid cells. It was certainly not increased as would be expected for recessive mutations, nor half the frequency as in the case of codominance. One might ask about the longevity of haploid frogs. Mezger-Freed informs us that haploid amphibians have a haploid disease, in which young animals become edematous and die. They thus do not live long enough to say whether or not they age at a normal rate. Harris and Mezger-Freed both considered stable phenotypic variation as a possible explanation for their results.

The conclusions of Harris and Mezger-Freed have been criticized however. Two important critiques are those of DeMars (1974) and Chasin and Urlaub (1974). The first deals with technical and statistical problems in measuring mutations and inconsistencies in the observations. In the second, Chasin induced mutants at the hypoxanthine phosphoribosyl transferase (HGPRT) locus on the X chromosome which confers resistance to 6-thioguanine in pseudo-tetraploid cells produced by hybridization. These were homozygous for HGPRT but heterozygous for the linked marker glucose-6-phosphate dehydrogenase (G6PD) Approximately half of the mutant selected concomitantly lost the wild type allele for G6PD as would be expected if both loss of one X chromosome and mutation on the remaining X had occurred. These chromosome segregations then lead to the diploid state in the

X chromosome, permitting the expression of a recessive mutation in an otherwise tetraploid cell.

We (Sinex, 1976) accepted stable phenotypic variants as a possible explanation of the Harris and Mezger-Freed results and then discussed the possibility that aging mutations are dominants rather than recessives. This view was based on experiments employing ionizing radiation. Doubling the number of chromosomes does not increase resistance to ionizing radiation ten thousand-fold. Lamb and Maynard Smith (1964) irradiated *Drosophila* and came to the conclusion that the mutant alleles induced by radiation may achieve the life shortening effects by combinations of dominant and recessive mutations especially in genes that regulate the activities of other genes. They concluded that induced heterozygote alleles are much more common than homozygotes and that even though the impairment produced in a heterozygote individual may be small, the overall effect of many heterozygote radiation injuries is much more important than the occasional homozygous injury. These conclusions of course are contrary to those made by Szilard (1959) in his paper on aging, who believed that only those aging hits which created homozygotes were effective. Maynard Smith's treatment of radiation injury in *Drosophila* should be applicable to injury in mammalian cells. Mutations produced in the course of time could be relatively minor impairments. In an ordinary genetic analysis they would be considered recessives, however in terms of their mild effects on fitness they are dominant in the sense that an effect is produced even though a single allele is affected.

Another way in which recessive genes can be expressed in somatic cells is through the parasexual cycle described by Martin and Sprague (1969). Here the number of chromosomes in a culture may go from 2n to 4n and back to 2n with two copies of a recessive chromosome now in a single cell. Or somatic chromosome pairing as observed by Heneen and Nichols (1972) may lead to expression of previously silent genes through somatic crossing over.

Direct investigation of mutation in late passage cells is difficult because the cloning of mutant cells requires division which is limited in such cells. Therefore the cells a gerontologist would most want to study are not accessible by conventional plating and cloning techniques. Mutation rate is being measured in earlier passage on cells undergoing rapid division under considerable stimulation from serum and a high oxygen tension.

We do believe that there is a somatic mutation component to aging and that this is most likely produced by the gap between the ability of a repair endonuclease to recognize a site of injury to DNA and the effect of the injury and ultimate mutation in producing a protein altered in function or quantity. Whether the primary mechanism is due to lack or reduction of a repair process or to the utilization of an error-prone repair mechanism as described by Strauss (1974) is not known. The placement of protein in replications of the chromatin of daughter cells may be more subject to error than base replication in DNA. We know very little about the process. A disturbing aspect of this approach is that enucleate frog eggs can receive tadpole nuclei and develop normally. This would suggest that at least for an egg it is possible to go back to the primary genetic information without consideration of any previous state of differentiation (Gurdon, 1973).

It is interesting to speculate on how repair of altered protein placement might occur in analogy with the action of excision endonuclease and post replication repair on DNA. One might postulate that the removal of misplaced protein might be the function of specific chromatin proteins. Repair of chromatin would require synthesis of chromatin protein by post mitotic cells. Very long-lived species would be unusually susceptible to time-dependent injury in their chromatin protein. Turnover of many of these proteins is slow. Spontaneous loss of amide nitrogen, oxidation of SH (Tas, 1976), spontaneous racemization of amino acids (Martin, et al., 1977), and other chemical changes could affect differentiation on a random basis. The spontaneous loss of amide nitrogen is known to occur in a number of proteins, the best example being crystallin, the protein in the lens of the eye (Sinex, 1960; Robinson, 1974; Palmer & Papaconstantinous, 1969). We hope to study amide nitrogen of chromatin protein as a function of age, to see if it might prove to be a factor in human senile dementia. With denaturation one might assume that specific interaction of chromatin protein with DNA would become weaker and nonspecific interaction stronger. Therefore in our view aging may be the result of two types of variation, one conventional type of mutation occurring in DNA, and the other due to altered arrangement or altered properties of chromatin protein. These are random. They may be random in two ways, in terms of the genes involved, and in terms of how long an effect persists in any single cells. However, the effects of many different random differentiations may lead to similar outcomes since cells can only express variations in a limited number of ways which can be visualized morphologically.

Little work has been done on the specific biochemical characteristics of late passage phenotypes. The paper of Mayne et al. (1977) is significant. They show that late passage chondrocyte cell cultures switch from the synthesis of cartilage specific type II collagen, to the synthesis of type I and type I trimer, more common types synthesized by skin fibroblasts. This switching is increased by the presence of BrdU. The switching is unusually sensitive to BrdU. The mechanism is not due to AT⇄GC transitions in DNA but is probably due to altered interaction between DNA and chromosomal protein.

There are more cells in the human body than there are people in the world. Microevolution within this society of cells would be expected unless extraordinarily well controlled. A certain permissiveness might be desirable to allow for adaptation to genetic variation in the individual. Somatic survival of a cell's progeny is not necessarily compatible with the survival of the organism. Cancer is the most obvious example. Both stable phenotypic variants and mutated cells may behave in a socially undesirable way within the society of cells which comprise the organism. They may make the wrong form of collagen, accumulate lipid, and propagate in the wrong places. Of the variant cells produced in aging not all may be adapted for survival or favorable propagation. Some are inferior, function poorly and are fated to die. Such cells would resemble those which appear in late passage from cultures of embryonic diploid human fibroblasts.

One way to visualize what we have been discussing is to imagine an aging tissue which is making hemoglobin in small amounts, and assume that hemoglobin making cells could be visualized microscopically. Would the tissue be pink (all cells exhibiting the same programmed property); would a few cells be red (suggesting mutation or stable phenotypic variation); or would the cells blink on and off (random phenotypic variation analogous to the gateing of mitosis)?

In conclusion, we believe that cellular aging may have multiple components including somatic mutation, the formation of phenotypic variants both stable and randomly fluctuating and the interaction of cells with hormones and other factors in their environment. Subsequent chapters will suggest experimental approaches for resolving some of these issues, and assigning to the various possible mechanisms their appropriate contributions to aging.

REFERENCES

Alexander, P., and Connell, D.I., 1960. Shortening of the life span of mice by irradiation with x-rays and treatment with radiomimetic chemicals. Radiation Res. 12:38-48.

Burnet, Sir MacF., 1974. INTRINSIC MUTAGENESIS: A GENETIC APPROACH TO AGING, Wiley, New York-Toronto.

Casarett, G.W., 1964. Similarities and contrasts between radiation and time pathology. In: ADVANCES IN GERONTOLOGICAL RESEARCH I, B.L. Strehler, ed., pp. 109-163, Academic Press.

Chasin, L.A. and Urlaub, G., 1975. Chromosome-wide event accompanies the expression of recessive mutations in tetraploid cells. Science 187:1091-1093.

Christiansen, G., and Griffith, J., 1977. Salt and divalent cations affect the flexible nature of the natural beaded chromatin structure. Nucleic Acids Res. 4:1837.

Clark, A.M., and Rubin, M.A., 1961. The modification by x-irradiation of the lifespan of haploid and diploids of the wasp, *Habrobracon sp.* Radiation Res. 15:244.

Curtis, Jr., H., 1966. BIOLOGICAL MECHANISMS OF AGING, Thomas, Springfield.

Curtis, J.M., and Miller, K., 1971. Chromosomal aberrations of liver cells of guinea pigs. J. Gerontol. 26:292-293.

DeMars, R., 1974. Resistance of cultured human fibroblasts and other cells to purine and pyrimidine analogues in relation to mutagenesis detection. Mutation Res. 24:335-364.

Gurdon, J.B., 1973-74. Gene expression in early animal development: The study of its control by the microinjection of amphibian eggs. Harvey Lectures 69:49-69.

Harris, M., 1971. Mutation rates in cells at different ploidy levels. J. Cell. Physiol. 78:177-184.

Hart, R.W. and Setlow, R.B., 1974. Correlation between deoxyribonucleic acid excision repair and life-span in a number of mammalian species. Proc. Nat. Acad. Sci. USA 71:2169-2173.

Hart, R.W. and Setlow, R.B., 1977. DNA repair in late passage human cells. Mechan. Aging and Devel. 5:67-77.

Heneen, W.K. and Nichols, W.W., 1972. Nonrandom arrangement of metaphase chromosomes in cultured cells of the Indian deer *Muntiacus muntjac*. Cytogenetics II:153-164.

Herrmann, R.L., Bach, M.D., Dowling, L. and Russell, A.P., 1975. Age-related changes in a spontaneous reassociating fraction of mouse DNA. Mech. Age. and Devel. 4:181-189.

Jacobs, P.A., Brunton, M., Court Brown, W.M., Doll, R. and Goldstein, H., 1963. Change of human chromosome count distributions with age: evidence for a sex difference. Nature, Lond., 197:1080.

Jacobs, P.A., Court Brown, W.M., and Doll, R., 1961. Distribution of human chromosome counts in relation to age. Nature, Lond., 191:1178.

Jarvik, L.F., Yen, F., and Moralishvili, E., 1974. Chromosome examinations in aging institutionalized women. Jrl. Gerontol. 29:269-2761

Kihlman, B.A., 1966. ACTIONS OF CHEMICALS ON DIVIDING CELLS, Prentice Hall, Englewood-Cliffs, N.J. 260 pp.

Kurtz, D.I., Russell, A.P., and Sinex, F.M., 1974. Multiple peaks in the derivative melting curve of chromatin from animals of varying age. Mech. Aging and Devel. 3:37-49.

Lamb, M.J. and Maynard Smith, J., 1964. Radiation and aging in insects. Exper. Gerontol. 1:11-20.

Martin, G.M., and Sprague, C.A., 1969. Parasexual cycle in cultivated human somatic cells. Science 166:761-763.

Masters, P.M., Bada, J.L. and Sigler, J.S., 1977. Aspartic acid racemization in the human lens during aging and in cataract formation. Nature 268:71-73.

Mattern, M.R. and Cerutti, P.A., 1975. Age-dependent excision repair of damaged thymine from γ-irradiated DNA by isolated nuclei from human fibroblasts. Nature 254:450-452.

Mayne, R., Vail, M., Mayne, P. and Metler, E.J., 1976. Changes in type of collagen synthesized in clones of chick chondrocytes grow and eventually lose division capacity. Proc. Nat. Acad. Sci. 73: 1674-78.

Mays, L., 1977. Abstracts of 7th Annual Meeting American Aging Association.

Mezger-Freed, L., 1972. Effect of ploidy and mutagens on bromodeoxyuridine resistance in haploid and diploid frog cells. Nature New Biol. 235:245-246.

O'Meara, A., and Herrman, R., 1927. A modified chromatin preparation displaying age related differences in salt dissociation and template ability. Biochem. Biophys. Acta., 269:419-429.

Palmer, W.G. and Papaconstantinou, J., 1969. Aging of α-crystallins during development of the lens. Proc. Nat. Acad. Sci. 64:404-10.

Robinson, A.B., 1974. Evaluation and the distribution of glutaminyl and asparaginyl residues in proteins. Proc. Nat. Acad. Sci. 71:885-888.

Sinex, F.M., 1976. Molecular genetics of aging. In: HANDBOOK ON THE BIOLOGY OF AGING, C. Finch and L. Hayflick, eds. Van Nostrand Reinhold, New York, pp. 37-62.

Sinex, F.M., 1974. The mutation theory of aging. In: THEORETICAL ASPECTS OF AGING, M. Rockstein, ed., pp. 23-31, Academic Press, New York.

Sinex, F.M., 1960. Aging and the liability of irreplaceable molecules. II amide groups of collagen. Jrl. Gerontol. 15:15-18.

Strauss, B.S., 1974. Repair of DNA in mammalian cells. Life Sciences 15:1685.

Strehler, B.L., 1959. Origin and comparison of the effects of time and high-energy radiations on living systems. The Quarterly Rev. of Biol. 34:117-142.

Szilard, L., 1959. On the nature of the aging process. Proc. Nat. Acad. Sci. 45:30-45.

Tas, S., 1976. Disulfide bonding in chromatin proteins with age and a suggested mechanism for aging and neoplasia. Exp. Gerontol. 11:17-24.

Upton, A.C., 1957. Ionizing radiation and the aging process, a review. Jrl. Gerontol. 12:306-313.

Upton, A.C., 1977. Pathobiology. In: HANDBOOK OF THE BIOLOGY OF AGING, C. Finch and L. Hayflick, eds. Van Nostrand Reinhold, New York, pp. 513-535.

SENESCENCE IN CELL CULTURE: AN ACCUMULATION OF ERRORS OR TERMINAL DIFFERENTIATION?

Vincent J. Cristofalo

The Wistar Institute of Anatomy and Biology

36th Street at Spruce, Philadelphia, Pa. 19104

Senescence in cell culture represents an ideal model system to examine a variety of questions related to cellular aging. The process of aging *in vitro* can be followed conveniently and precisely by measurement of the saturation density of the culture (1), the percentage of clones that can proliferate (2) or the percentage of cells incorporating ^{3}H-thymidine under a given standard set of conditions (3). In addition, other physiologic changes which parallel the losses in proliferative capacity such as the increase in lysosomes and lysosomal enzyme activity (4-6) can be conveniently studied using proliferative activity as an independent measure of population age. In these cultures, most, if not all, environmental factors can be controlled. Thus, we can design experiments to determine the answer to fundamental questions at the cellular level which could not be answered easily *in vivo*.

It is important to point out, however, that the theoretical basis for aging in a continually proliferating cell culture may be unrelated to that of fixed postmitotic cells either *in vivo* or in culture. For the purposes of this discussion, then, I would like to restrict my remarks to the data available from proliferating cell cultures in evaluating the theoretical mechanisms of aging.

Perhaps the most active area of investigation addressing the role of errors in cell aging has involved the work of Holliday, Tarrant, Lewis and their colleagues (7-9) in using cell cultures to test the error theory of Orgel as it applies to aging (10,11). Orgel's theory states that aging is the result of a random accumulation of errors that occur during protein synthesis. Most error-containing abnormal proteins would be diluted out through the continual turnover and replacement by newly synthesized error-free molecules. However, errors that occur in those proteins (enzymes) involved in the protein synthesizing machinery could be self-perpetuating in that new errors would be continually produced. Eventually, a cascade of errors could be produced which would result in an irreversible breakdown in the fidelity of the protein synthesizing machinery and the ultimate functional failure and death of the cell.

The approach of these workers has been to measure the presence of errors in young and old cells in one or two specific proteins using such parameters as changes in heat inactivation kinetics and an increase in enzymatically inactive molecules which immunologically cross react. They have reported up to 25% abnormal thermolabile glucose-6-phosphate dehydrogenase and 6-phosphogluconate dehydrogenase in senescent MRC-5 fibroblast-like cells. Similarly, they have found increases in enzymatically unreactive lactate dehydrogenase in old cultures. Goldstein and his coworkers (12,13) have found age-associated increases in heat labile glucose-6-phosphate dehydrogenase, 6-phosphogluconate dehydrogenase and hypoxanthine-guanine phosphoribyl transferase from normal skin cultures. Both Holliday *et al*. (9) and Goldstein and his coworkers (12-14) have found this same kind of increase in cells derived from patients with Werner's syndrome. Similarly, Goldstein and Moerman (14) have found increases in heat labile enzyme in cell cultures derived from progeric individuals.

It is important to point out that increases in aberrant proteins in senescent cultures does not specify missynthesis since variations in post-synthetic modifications of proteins can also yield the same results; nonetheless, these findings are consistent with the notion that increased errors in protein synthesis could be responsible for the functional failures observed in aging *in vitro*.

In a series of studies directly paralleling the studies of Holliday and his coworkers, Pendergrass *et al*. (15) have examined human male infant skin fibroblast-like cultures for the appearance of thermolabile and immunologically active but enzymatically inactive glucose-6-phosphate dehydrogenase during *in vitro* aging. Their results contrast sharply with those of Holliday and his coworkers who used MRC-5 cultures which were derived from male fetal lung (7).

In these latter studies, the thermostability curves of glucose-6-phosphate dehydrogenase at 10 and 55 doublings were essentially identical with no evidence for the appearance of a thermolabile component in the senescent culture. Similarly, when enzyme preparations were titrated with antiserum prepared against purified glucose-6-phosphate dehydrogenase, the titration curves were identical.

Fortuitously, during the course of the studies, these workers became aware of an individual who had mutant glucose-6-phosphate dehydrogenase which had a thermolabile component and thus cells from this person were used as a valuable positive control, showing that the techniques being used were adequate to determine the heat labile enzyme if it were present.

In another series of studies, the Gershons and their co-workers (16) using young and senescent mouse fibroblast-like cells could not detect differences in the thermostability of the enzyme aldolase nor changes in the immunologically reactive but enzymatically inactive enzyme during aging.

Other findings which are difficult to reconcile with the error catastrophe hypothesis in human diploid cells have been presented by Holland (17) and Tomkins (18) independently. They have compared and found no differences between the kinetics of production of infectious virus and the specific infectivity of virus grown in senescent and young human cells. In addition, the kinetics of thermal inactivation of vesicular stomatitis virus and the mutation rate of poliovirus-induced replicase in old human diploid cell cultures were not significantly altered during virus replication in senescent cells. This would indicate that the fidelity of the cellular transcriptional and translational machinery that is necessary for production of normal viruses is maintained in old cultures.

The basis for the differences among the results of these workers is obscure and may be due to various technical differences in the procedures. As mentioned above, the presence of error-containing proteins does not define errors in protein synthesis. Various differences in post-translational processing could result in the protein differences observed. For example, the rate and extent of degradation of the proteins could be different in cells cultured differently or with different homogenization procedures. It has been reported that lysosomal enzymes increase in aging cells and that the mobility of these enzymes is also increased (4,19). Possibly lysosomal proteases have different access to the enzymes being studied in the different procedures used.

To take a negative view, one could argue that implicit in the Orgel theory is the assumption that the rate of error production is much higher than the rate of cell division so that selection does not take place. However, there is no evidence to support this assumption. In addition, although protein missynthesis may play a role in aging in fixed post-mitotic tissues in vivo, it is difficult to reconcile the error hypothesis as it applies to aging in continually proliferating systems with the fact that, in these systems, cells containing damaged, survival-compromising protein molecules would be selected against and diluted out by the routine procedures of serial subcultivation.

Further arguments can be derived from some very simple experiments of ours addressing the question of how the error theory might operate in vitro (20). If amino acid analogues were included in the culture medium and shown to be incorporated into proteins, then their presence should result in an increased incidence of errors and a shortened lifespan for the cultures. For these experiments, we used two analogues in combination. Initially we determined the dose-sensitivity characteristics for the combination of p-fluorophenylalanine and ethionine on WI-38 cells during a single growth cycle. At 10^{-6} μM concentrations of each of these two analogues, they can be shown to be incorporated into proteins by radioactivity measurements, but appear to have no effect on the growth rate. Higher concentrations were clearly inhibitory to growth. With 10^{-6} μM analogues present in the medium throughout the lifespan of the culture, there were no differences in cell yield/subcultivation, nor in the continued presence of these analogues was there a shortening of the lifespan.

In the companion experiment, cultures were seeded into medium containing 10^7- and 10^8-fold higher concentrations of the analogues and maintained for one week. At these concentrations, there was gross toxicity and proliferation was severely inhibited. After one week, the analogue-containing medium was renewed and fresh medium was added and the cultures allowed to proliferate; here, the cultures appeared to recover completely, as evidenced by the fact that the growth rate and the lifespan were essentially the same as that of the controls. Thus neither the continuous presence of low levels of amino acid analogues nor the acute exposure to very toxic levels shortened the lifespan.

To espouse error theory, as an explanation for aging in cell culture, one needs to explain why transformed cells under the same growth conditions are not subject to errors. The usual explanations proposed deal with the assumption that some fundamental metabolic pathways are different in transformed cells. There is no evidence to support this notion. However, if it were true, one would at least expect a different degree of sensitivity of normal and transformed cells to a given analogue concentration. In our study of dose-sensitivity relationships of WI-38 and SV-40-transformed WI-38 cells over a wide range of concentrations of these same analogues, we found no real differences.

These data, then, do not support the Orgel error hypothesis and, in fact, do indicate that selection takes place in the population for the cells that are most "fit."

Another area that we have investigated is that of the role of free radical reactions in aging. A number of authors have proposed that normal cellular function is compromised by free radical reactions involving both the nucleus in, for example, somatic mutation as well as in lipid peroxidation in various membranes.

The molecule that most commonly generates free radicals is oxygen and Dr. Balin in our laboratory has been studying the effects of reduced oxygen on lifespan in WI-38 cells (21,22).

We found that when cultures were carried continuously for 25 weeks at different partial pressures of oxygen, there was no difference in the lifespan between cultures grown at atmospheric partial pressures of oxygen and at pressures reduced to approximately 1/3 or 1/5 the ambient amount. The fact that these reduced partial pressures of oxygen did not extend the lifespan suggests that free radical reactions do not play a really significant role in limiting culture lifespan. This conclusion is further supported by our recent studies with the free radical trapping agent vitamin E.

Packer and Smith (23) have suggested that the antioxidant dl α tocopherol or vitamin E extended the _in vitro_ lifespan of WI-38 cells by a factor of about 2. This of course would fit in very well with the free-radical theory of aging, however, we have been unable to reproduce that result. Dr. Balin in our laboratory has carried out extensive experiments in which vitamin E at two concentrations was homogenized into tissue culture media according to the method of Packer and Smith (23). Cells were also grown in homogenized media without vitamin E. Experiments were carried out in air containing 5% CO_2 and at gas mixtures giving low, ambient, and high partial pressures of oxygen. Vitamin E had no effect on cell lifespan under any of these conditions. [In an addendum to their paper, Packer and Smith (23) indicated that they were unable to duplicate their results when the serum lot was changed.]

In concluding, I would like to briefly consider an alternative view of cell aging in cultures (24-26), viz that the observed cellular degeneration is the result of a sequential process of a terminal differentiation such as exists for hematopoietic cells, for example. Diploid cell populations in culture undergo a regular decline in proliferative activity which appears to be the result of individual cells in the population proceeding asynchronously from a rapidly cycling state to a series of more slowly cycling states, and finally to a state of arrest or at least apparent arrest within the limits of the methods used to detect cell cycling (3,27,28). These changes are accompanied by several well-defined physiological changes. For example, the nonproliferating state is associated with increased size (29-31), increased content of secondary lysosomes (5,6) and a decline in chromatin template activity (32). These changes would seem to be consistent with changes in gene expression rather than with the effects of random errors and tend to support the alternative view of a programmed terminal differentiation being responsible for aging in cell culture.

This work was supported in part by USPHS research grants AG 00378 from the National Institute of Aging and RR 05540 from the Division of Research Resources.

1. Hayflick, L. and Moorhead, P.S. 1961. The serial cultivation of human diploid cell strains. Exp. Cell Res. 25: 585.

2. Merz, G.S. and Ross, J.D. 1969. Viability of human diploid cells as a function of in vitro age. J. Cell Physiol. 74: 219.

3. Cristofalo, V.J. and Sharf, B.B. 1973. Cellular senescence and DNA synthesis. Thymidine incorporation as a measure of population age in human diploid cells. Exp. Cell Res. 76: 419.

4. Cristofalo, V.J., Parris, N. and Kritchevsky, D. 1967. Enzyme activity during the growth and aging of human cells in vitro. J. Cell Physiol. 69: 263.

5. Robbins, E., Levine, E.M. and Eagle, H. 1970. Morphologic changes accompanying senescence of cultured human diploid cells. J.Exp. Med. 131: 1211.

6. Lipetz, J. and Cristofalo, V.J. 1972. Ultrastructural changes accompanying the aging of human diploid cells in culture. J. Ultrastruct. Res. 39: 43.

7. Holliday, R. and Tarrant, G.M. 1972. Altered enzymes in ageing human fibroblasts. Nature 238: 26.

8. Lewis, C.M. and Tarrant, G.M. 1972. Error theory and ageing in human diploid fibroblasts. Nature 239: 316.

9. Holliday, R., Porterfield, J.S. and Gibbs, D.D. 1974. Premature ageing and occurrence of altered enzyme in Werner's syndrome fibroblasts. Nature 248: 762.

10. Orgel, L.E. 1963. The maintenance of the accuracy of protein synthesis and its relevance to aging. Proc. Nat. Acad. Sci. USA 49: 517.

11. Orgel, L.E. 1973. Ageing of clones of mammalian cells. Nature 243: 441.

12. Goldstein, S. and Singal, D.P. 1974. Alteration of fibroblast gene products *in vitro* from a subject with Werner's syndrome. Nature (Lond.) 251: 719.

13. Goldstein, S. and Moerman, E.J. 1976. Defective proteins in normal and abnormal human fibroblasts during aging *in vitro*. Interdiscipl. Topics Geront. 10: 24.

14. Goldstein, S. and Moerman, E.J. 1975. Heat-labile enzymes in skin fibroblasts from subjects with progeria. New Engl. J. Med. 292: 1306.

15. Pendergrass, W.R., Martin, G.M. and Bornstein, P. 1975. Evidence contrary to the protein error hypothesis for in vitro senescence. J. Cell. Physiol. 87: 3.

16. Danot, M., Gershon, H. and Gershon, D. 1975. The lack of altered enzyme molecules in "senescent" mouse embryo fibroblasts in culture. Mech. Ageing Develop. 4: 289.

17. Holland, J.J., Kohne, D. and Doyle, M.V. 1973. Analysis of virus replication in ageing human fibroblasts cultures. Nature 245: 316.

18. Tomkins, G.C., Stanbridge, E.J. and Hayflick, L. 1974. Viral probes of aging in the human diploid cell strain WI-38. Proc. Soc. Exp. Biol. Med. 146: 385.

19. Cristofalo, V.J. and Kabakjian, J. 1975. Lysosomal enzymes and aging *in vitro*: subcellular enzyme distribution and effect of hydrocortisone on cell lifespan. Mech. Ageing Develop. 4: 19.

20. Ryan, J.M., Duda, G. and Cristofalo, V.J. 1974. Error accumulation and aging in human diploid cells. J.

Gerontol. 29: 616.

21. Balin, A.K., Goodman, D.B.P., Rasmussen, H. and Cristofalo, V.J. 1976. The effect of oxygen tension on the growth and metabolism of WI-38 cells. J. Cell. Physiol. 89: 235.

22. Balin, A.K., Goodman, D.B.P., Rasmussen, H. and Cristofalo, V.J. 1976. The effect of oxygen and vitamin E on the lifespan of human diploid cells in vitro. Submitted for publication.

23. Packer, L. and Smith, J.R. 1974. Extension of the lifespan of cultured normal human cells by Vitamin E. Proc. Nat. Acad. Sci. USA 71: 4763.

24. Cristofalo, V.J. 1972. Animal cell cultures as a model system for the study of aging. Adv. Gerontol. Res. 4: 45.

25. Martin, G.M., Sprague, C.A., Norwood, T.H. and Pendergrass, W.R. 1974. Clonal selection, attenuation and differentiation in an in vitro model of hyperplasia. Am. J. Path. 74: 137.

26. Martin, G.M., Sprague, C.A., Norwood, T.H., Pendergrass, W.R., Bornstein, P., Hoehn, H. and Arend, W.P. 1975. Do hyperplastoid cell lines "differentiate themselves to death" in cell impairment. In Aging and Development. V.J. Cristofalo and E. Holeckova, eds. Plenum Press, N.Y. p. 67.

27. Macieria-Coelho, A., Loria, E. and Berumen, L. 1975. Relationship between cell kinetic changes and metabolic events during cell senescence in vitro in cell impairment. V.J. Cristofalo and E. Holeckova, eds. Plenum Press, N.Y. p. 51.

28. Absher, P.M., Absher, R.G. and Barnes, W.D. 1974. Genealogies of clones of diploid fibroblasts: cinemicrophotographic observations of cell division patterns in relation to population age. Exp. Cell Res. 88: 95.

29. Cristofalo, V.J. and Kritchevsky, D. 1970. Cell size and nucleic acid content in the diploid human cell line WI-38 during aging. Medicinia Experimentalis 19: 313.

30. Bowman, P.D., Meek, R.L. and Daniel, C.W. 1975. Aging of human fibroblasts in vitro. Correlations between DNA synthetic ability and cell size. Exp. Cell Res. 93: 184.

31. Mitsui, Y. and Schneider, E.L. 1976. Relationship between cell replication and volume in senescent human diploid fibroblasts. Mech. Ageing Develop. 5: 45.

32. Ryan, J.M. and Cristofalo, V.J. 1975. Chromatin template activity during aging in WI-38 cells. Exp. Cell Res. 90: 456.

SYNKARYON AND HETEROKARYON ANALYSES OF CLONAL SENESCENCE

Thomas H. Norwood, Holger Hoehn, Andrew O. Martinez,
George M. Martin

Department of Pathology, University of Washington
Seattle, Washington 98195

INTRODUCTION

The observation of Hayflick and Moorhead in 1961 that human diploid cells have a limited replicative *in vitro* life span (clonal senescence) (1) has been confirmed in many laboratories throughout the world. In addition to their predictable proliferative behavior, they are remarkably resistant to spontaneous or induced transformation to immortal cell lines.

The significance of clonal senescence to *in vivo* aging remains controversial (2), although four lines of evidence suggest a relationship to *in vivo* aging: 1) the replicative life span of cultures are inversely related to the ages of the donors (3-6); 2) the growth potential of somatic cells is a function of the tissue of origin (5,7,8); 3) a diminished growth potential has been observed for cultivated cells derived from individuals with genetically determined syndromes which display features of premature aging (5,9) and 4) some evidence suggests a correlation between the replicative life span of cultured cells and the life span of donor species (10). Detailed considerations of the relevance of this system as a model for *in vivo* aging have been reviewed in previous publications (10, 11). For the purposes of this discussion, it will be assumed that this system is a valid model for the study of aspects of *in vivo* senescence, although it is clear that much more work needs to be done.

The basic mechanism(s) involved in the loss of replicative potential of cultured human cells remains elusive, in spite of extensive investigations by a number of laboratories during the last decade. Hypotheses fall into two broad categories; these categories are not mutually exclusive: 1) many investigators have suggested that injurious stochastic events are the primary cause of clonal senescence, with or without genomic modulation. The protein error catastrophe theory of Leslie Orgel (12) is an important example; it has had a profound effect on aging research in both in vivo and in vitro systems. Orgel postulates a cascade of abnormal cell proteins resulting from mistakes in transcription and/or translation, the rate and character of which could be modulated by scavenger enzymes (13,14). Other stochastic models of senescence implicate somatic mutational events (15,16) or intra-cellular free radical formation (17,18) as the primary causes of senescence. Implicit in all of these models is that genetically controlled corrective mechanisms are present and may be important in determining the longevity of a given species (or cultivated cell type). 2) Other authors have emphasized the possibility of a genetic program controlling the proliferative capacity of dividing cell populations in vivo or in vitro. It has been suggested that cultivated diploid mammalian cells undergo a series of genetically controlled steps, the end result of which is a post-replicative fully differentiated state (11,19,20). Clearly, however, secondary injurious, stochastic events could occur in the noncycling cell resulting in their ultimate demise. Related to the second class of theory concerning clonal senescence is that of a "running out of genetic program" or a kind of "end of the tape" theory. Since simple clock models of clonal senescence have virtually been ruled out (19,21), such a theory must incorporate elements of one or both of the above models.

In an attempt to determine which, if any, of the above theories might best explain the mechanism of clonal senescence, we initiated studies of the complementary behavior (i.e., the dominance or recessivity) of the replicative behavior of senescent human diploid cells following fusion to actively proliferating cell types.

We have employed three basic types of experiments in the genetic analysis of senescent clones of human diploid cells: 1) heterokaryon studies, in which nuclei are separate but in a common cytoplasm; 2) heteroplasmosis (22), in which one parent consists of anucleate cytoplasm or a nucleus with residual remnant of cytoplasm; 3) synkaryon studies, in which nuclear as well as cytoplasmic fusions result in proliferating clones.

Heterokaryon Studies

The rationale for our initial heterokaryon experiments was based on the differentiation hypothesis described above. Professor Henry Harris and his associates (23,24) were able to demonstrate reinitiation of nuclear macromolecular synthesis in cells of known differentiated function (e.g., chick erythrocytes) after heterokaryon formation with actively proliferating cells; in effect, dominance of the "undifferentiated," actively dividing cell. We reasoned that if postmitotic human diploid cells were, in fact, differentiated cells, we should be able to demonstrate evidence of nuclear DNA synthesis under similar experimental conditions. In our first series of experiments we induced heterokaryon formation between young and late passage human diploid cells via cell fusion with Sendai virus (25). The nucleus of the young parental cell was prelabeled with [^{14}C] thymidine and the cytoplasm of the senescent partner was prelabeled with [^{3}H] methionine. These prelabels, combined with double layer autoradiography (26), allowed us to identify the parental origins of bi- and multi-nucleated cells. Immediately after cell fusion, 3 serial 24 hour pulses of [^{3}H] thymidine were initiated in order to monitor nuclear DNA synthesis. The results (Figure 1) indicated dominance of the senescent phenotype with respect to nuclear incorporation of [^{3}H] thymidine in both homologous and isologous crosses. There was no evidence of nuclear DNA synthesis in senescent nuclei; moreover, it was markedly inhibited in the young nuclei.

This observation certainly did not support the differentiation hypothesis and could be interpreted as supporting a stochastic theory, which predicts the accumulation of toxic substances. However, in Harris' experiments, with one exception (29), HeLa or Ehrlich ascites cells were used as the actively proliferating parent. These cell lines typify a class of cultured somatic cells to which we have applied the term "neoplastoid," the salient characteristics of which include an apparently unlimited replicative potential, aneuploidy and, under appropriate conditions, tumorigenicity (Table I). By contrast, human diploid cells typify a class we have called "hyperplastoid;" these cells are of limited replicative potential, remain diploid throughout their *in vitro* life span and are apparently nontumorigenic.

These heterokaryon experiments were repeated using cells from two neoplastoid cell lines (HeLa and SV-40 transformed human fibroblasts) as the actively proliferating parents (28). In contrast to the previous experiments, there was evidence of reinitiation of DNA synthesis in the senescent nuclei when crossed with either HeLa or SV-40 (Figures 2 and 3). There was, however, a difference in the kinetics of labeling, in that there was a decline in the extent of [^{3}H] thymidine labeling in the HeLa - old cross, not observed in the experiment with SV-40. We interpreted these results as evidence in support of the terminal differentiation hypothesis but clearly could not rule out alternative interpretations. We also suggested that elaboration of a specific repressor(s) by the senescent cells, specifically inhibitory to their own stem cells, could explain the apparent recessive behavior of low passage diploid cells in this system.

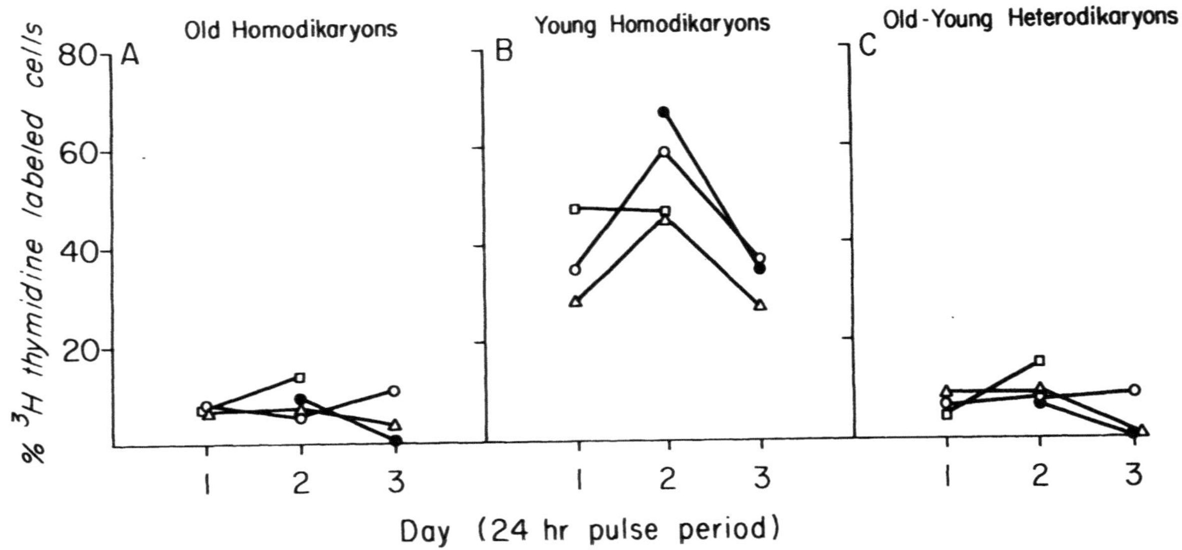

Figure 1: [^{3}H] thymidine labeling indices in (A) old homodikaryons, (B) young homodikaryons and (C) old-young heterodikaryons. After Norwood, et al. (25).

Table I. A Comparison of Some Properties of Hyperplastoid Cell Lines with Neoplastoid Cell Lines

Property	Hyperplastoid Cell Lines	Neoplastoid Cell Lines
Prototype Cultures	Human skin "fibroblasts" and WI-38	HeLa and L cells
Origin	Normal and neoplastoid stroma	Normal (ex-lymphocytes) and neoplastic cells
Replicative life span	Finite	Apparently infinite
Karyotype	Same as donor's normal somatic cells	Usually different from donor's normal somatic cells
Tumor assay	Negative	Sometimes positive
Growth in suspension culture	No	Frequently
Saturation density	Comparatively low	Comparatively high

Another interpretation of the experiments described above would be that neoplastoid cells possess extremely efficient error correcting mechanisms. To test this alternative we attempted to create an artificial protein error catastrophe by exposure of young actively dividing diploid cells to a combination of amino acid analogues known to be incorporated into cellular proteins. The duration and concentration of the analogue treatments allowed for survival of a fraction of nonreplicating cells. The analogue-treated cells were then fused to nontreated low passage diploid cells and to HeLa cells according to the protocol described above. We were somewhat surprised to find that the nuclear [^{3}H] thymidine labeling pattern in the heterokaryons observed in these experiments was, with minor exceptions, virtually identical to that observed in crosses with senescent cells. There are at least two plausible interpretations of these heterokaryon results: 1) the accumulation of abnormal proteins (error catastrophe) is the only, or at least a paramount, cause of clonal senescence _in vitro_ and 2) this experimental system is nondiscriminating, i.e., all or most nondividing cells behave in a similar fashion, regardless of the causal mechanism.

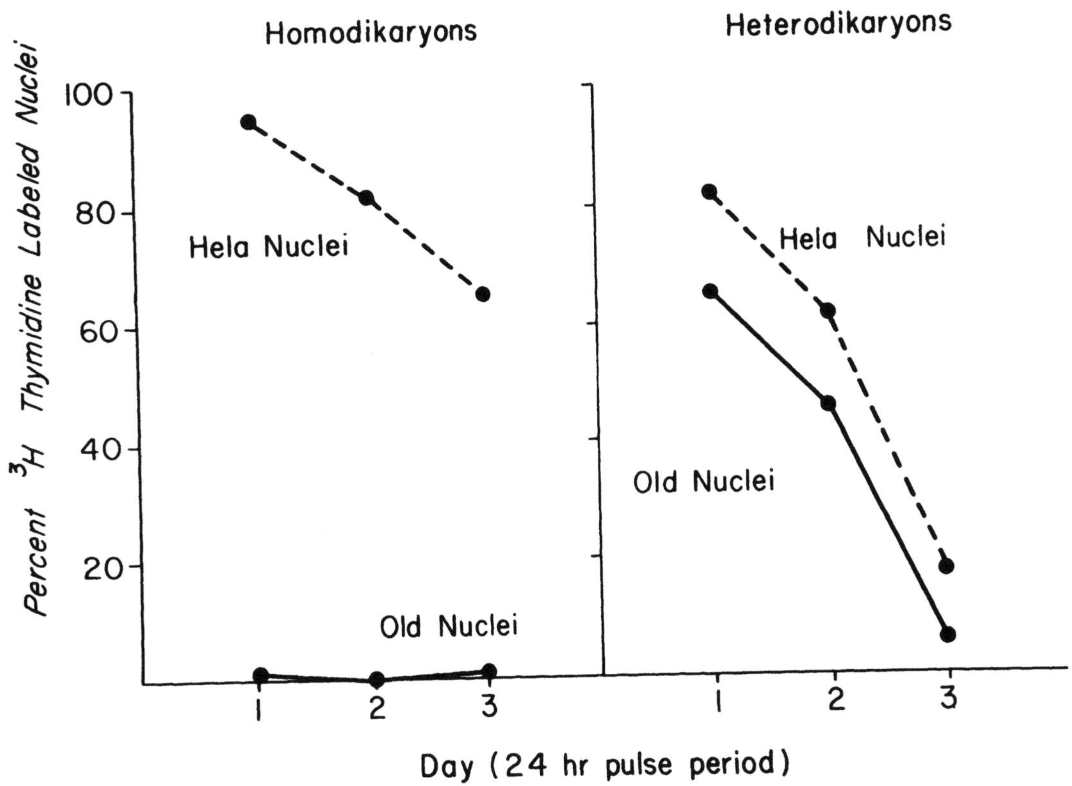

Figure 2. [^{3}H] thymidine labeling indices of nuclei within homodikaryons and by Sendai virus mediated fusion of senescent human skin fibroblasts and HeLa cells. After Norwood, et al. (28); reproduced by permission of the publisher.

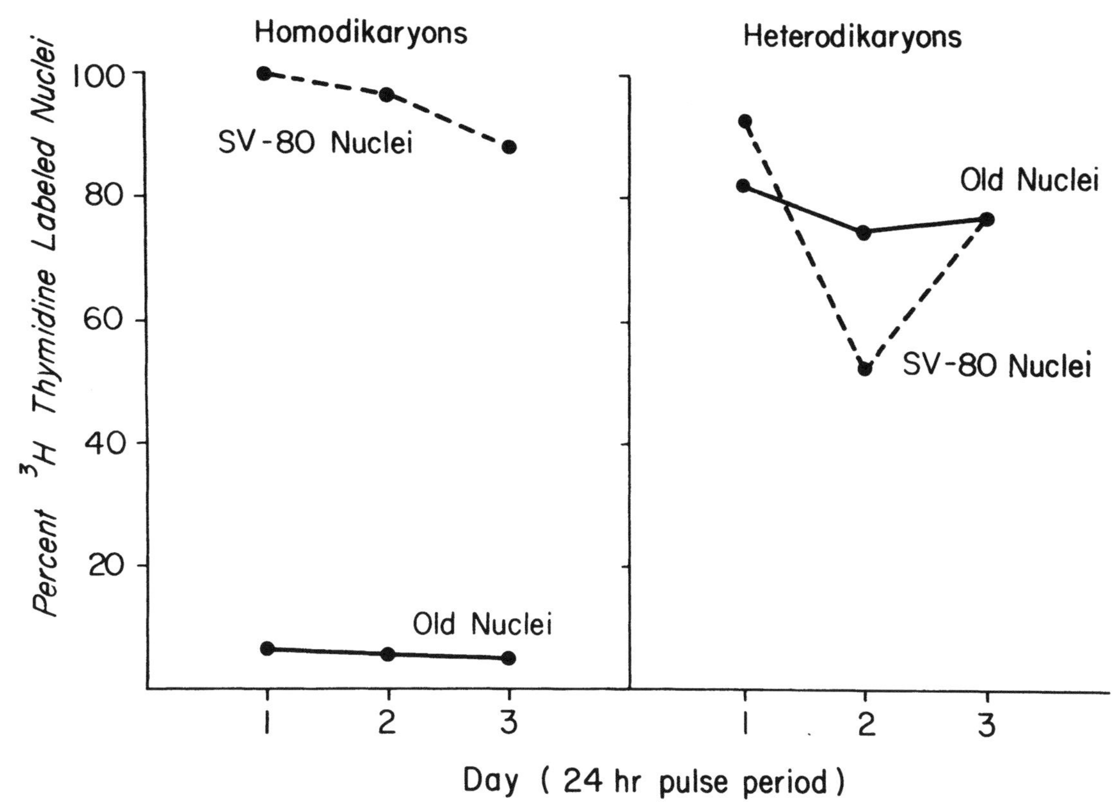

Figure 3. [^{3}H] thymidine labeling indices of nuclei within homodikaryons and heterokaryons obtained by Sendai virus mediated fusion of senescent human skin fibroblasts and a line of human skin fibroblasts which had been transformed by SV-40 virus. After Norwood, et al. (28); reproduced by permission of the publisher.

Knowledge of the biologic behavior of the cells following the analogue treatment is, of course, important for the interpretation of these experiments. Morphologically they appear quite distinct from the senescent phenotype. Long term observations of treated cells reveal clonal recovery in cultures. However, the majority of surviving cells show no evidence of recovery of cell cycle function. These surviving clones could be derived from: 1) resistant variants within the population; 2) cells which somehow escaped significant injury or 3) injured cells which recovered, indicating a reversible type of injury. Studies are underway to determine if these clones display increased resistance to prolonged exposure to the same amino acid analogues. Although incomplete, these observations indicate that the phenotype of the analogue treated cells and the senescent cells are different. This suggests that this system does not discriminate between different causes of loss of replicative function. As will be discussed below, other studies are underway which will hopefully bring into sharper focus the significance of the heterokaryon analysis of clonal senescence.*

Heteroplasmosis Studies

Mass enucleation of cells by means of exposure to cytochalasin B and centrifugation is now an established technique (29,30). It would be of obvious interest to determine if the ability to induce thymidine incorporation in senescent nuclei is retained by isolated cytoplasms and/or nuclei of neoplastoid cells. To this end, we have modified a technique (31) for enucleation of cells by utilizing Corex glass centrifuge tubes (Corning Glassworks, Corning, New York) for both growth and centrifugation of the cell populations. The initial studies concentrated on the fusion of HeLa cytoplasts with intact senescent cells using Sendai virus. In these experiments the HeLa cells were prelabeled with [^{3}H] methionine (cytoplasmic label) and [^{14}C] thymidine (nuclear label). The senescent cells displaying a cytoplasmic label, after cell fusion, were assumed to have fused with HeLa cytoplasm. HeLa cells which had escaped enucleation could be identified by the nuclear label. The results of a representative experiment are shown in Table II. The senescent cells with a cytoplasmic label consistently displayed a 2-3 fold increase in the percent of thymidine labeled cells, as compared to those without cytoplasmic label. This is in contrast to intact HeLa cells, which induce a 10-fold increase in labeled senescent nuclei.

*A manuscript describing in detail the materials and methods of procedure is currently in preparation.

Table II. Percent [^{3}H] thymidine labeled nuclei in mononucleate, senescent human diploid cells fused with cytoplasms from enucleated HeLa cells. The senescent cells displaying cytoplasmic label (i.e., silver grains over the cytoplasm) were assumed to have fused with the HeLa cytoplasm, which had been heavily prelabeled with [^{3}H] methionine. [1]250-500 cells counted; [2]25-100 cells counted; [3]senescent human diploid cells exposed to Sendai virus but not co-cultivated with HeLa cytoplasm.

		[^{3}H] thymidine pulse 0-24 hrs. after fusion		[^{3}H] thymidine pulse 24-48 hrs. after fusion	
Exp.	Slide No.	A[1]	B[2]	A[1]	B[2]
a	1	5.1	6.0	8.3	20.3
	2	5.1	9.0	8.0	15.7
	3	3.6	12.3	8.0	14.4
	4	0.78	8.0	6.9	14.6
	Mean	3.6	8.8	7.8	16.3
	Control[3]	4.2		5.7	
b	1	2.5	22.2*	2.1	---
	2	4.5	12.2	4.2	8.3
	Mean	3.5	17.2	3.2	8.3
	Control[3]	3.3		6.1	

*Only 9 cells counted
A Cells without cytoplasmic label
B Cells with cytoplasmic label

In more recent experiments, we have used polyethylene glycol to induce fusion between HeLa cytoplasts and senescent cells. This agent has been demonstrated to more efficiently and reproducibly fuse mammalian somatic cells (32,33). The fusion efficiency can be further enhanced by the addition of dimethylsulfoxide (34). In studies using this method of cell fusion, the number of senescent cells with cytoplasmic label and the density of label were far greater than those previously observed. As shown in Table III, the degree of stimulation of [^{3}H] thymidine incorporation in cells with cytoplasmic label was almost identical to that observed in the previous study. By day 3, the thymidine labeling indices were similar. This could be the result of dilution of the cytoplasmic label due to protein turnover; however, the fraction of identifiable cells with cytoplasmic label remained approximately constant, suggesting that this was not the case. Since control studies have not been

Table III. Percent [^{3}H] thymidine labeled nuclei in mononucleate senescent human diploid cells fused with cytoplasms from enucleated HeLa cells using polyethylene glycol to induce cell fusion. The experimental protocol is as described for Table II. [1]150-350 cells counted; [2]150-400 cells counted; [3]1000-1500 cells counted in senescent human diploid cells exposed to polyethylene glycol but not co-cultivated with HeLa cytoplasm.

	[^{3}H] thymidine pulse 0-24 hrs.		[^{3}H] thymidine pulse 24-48 hrs.		[^{3}H] thymidine pulse 48-72 hrs.	
Exp.	A[1]	B[2]	A[1]	B[2]	A[1]	B[2]
a	1.2	3.8	1.4	2.7	2.6	2.5
b	1.4	3.2	2.6	4.7	2.8	2.4
Mean	1.3	3.5	2.0	3.7	2.7	2.5
Control[3]	1.8		2.7		2.2	

A Cells without cytoplasmic label
B Cells with cytoplasmic label

completed, the cause of the diminished stimulatory capacity of isolated cytoplasms in comparison to intact HeLa cells is unclear. Intact HeLa cells exposed to cytochalasin B for a period equal to that required for our enucleation procedure showed no loss of stimulatory function. The use of trypsin for harvesting the cytoplasms could result in removal of proteolytic alteration of cell surface proteins which are essential for stimulation and cannot be regenerated in the enucleate cell. Experiments in which cytoplasts have been mechanically dislodged from the Corex tubes and fused to the senescent cells are currently underway. If this control is negative we shall then test the stimulatory capacity of intact cells which have been exposed to cytochalasin B and suboptimum centrifugation. It is, of course, possible that HeLa nuclei contain a high concentration of molecules responsible for the reinitiation of DNA synthesis in senescent fibroblasts.

Synkaryon Studies

Up to this point we have been discussing the interaction of senescent cells with respect to the induction of nuclear [^{3}H] thymidine incorporation, the nature of which is far from clear (to be discussed below). Thus, the relevance of these observations to normal cell cycle function remains to be established. The ability, or lack thereof, to form viable, hybrid cells with senescent cells (synkaryons) is one approach to this problem. The result of such studies published thus far correlate very well with our observations in the heterokaryon experiments; Littlefield (35) was able to detect no complementation between different pairs of senescent fibroblast culture and between senescent and young human diploid cells. On the other hand, Croce, et al. (36), and Goldstein and Lin (37) observed viable hybrid clones derived from crosses between senescent human diploid cells and neoplastoid cells, although these clones could have derived from a residual population of replicating cells in the senescent cultures rather than from truely postmitotic cells. Studies with human diploid cells are hampered by the lack of selective markers, as evidenced by the fact that no one has ever before isolated proliferating hybrids from such cell populations. To this end, we have developed a nonselective method for the isolation and propagation of hybrid clones of human diploid cells (38). A technique which proved useful was the visual comparison of the lengths of the equatorial plates of mitotic metaphases of putative tetraploid clones with control diploids growing in the same culture dishes (39). We could then screen presumptive hybrid clones by using rapid electrophoretic screening for a glucose-6-phosphate dehydrogenase (G6PD) polymorphic marker. We have attempted three different types of fusion experiments so far: 1) fusions between two different male strains, one G6PD A and one G6PD B; 2) fusions within a genetically mosaic G6PD A/B heterozygotic female fibroblast line and 3) fusions between fibroblasts from a G6PD A male and a G6PD B female with Werner's syndrome, a strain which is of exceedingly low growth potential (5). Approximately one hybrid clone per thousand was observed in the first two types of fusions; therefore, one would have expected approximately eight clones from among the more than 8,000 clones screened in the crosses between the G6PD A male and the Werner's syndrome patient. So far we have found none, suggesting dominance of the senescent phenotype in Werner's syndrome in this system. In Table IV is shown the summary of the life histories of the hybrid tetraploids and hybrid mixoploid clones isolated from the first two types of experiments (Table IV). Based on the still very limited numbers of observations, our initial conclusion is that their cumulative population doublings are similar to those diploid controls which are big enough to be transferred and passaged. Certainly none of these clones displayed greater proliferative capacity than those reported in previous cloning studies with human diploid cells (40).

DISCUSSION

At this time both heterokaryon and synkaryon studies indicate that senescent human diploid cells are dominant to young, dividing presenescent human cells, but are recessive to at least some neoplastoid cell lines. As state above, we have concluded that these results are consistent with a differentiation hypothesis of clonal senescence. To explain the apparent recessive behavior of

Table IV. In Vitro Life Histories of 5 Hybrid Isolates

Clone #	Percentage Tetraploid Cells Initial	Final	Cumulative Population Doubling
1	100	100	35.3
4	100	100	46.6
5	95	100	33.9
6	42	15	31.9
8	100	100	<25.0

the low passage cells we postulated repressor(s) (3) that specifically inhibited proliferative function in the stem cells. However, the heterokaryon experiments using analogue treated fibroblasts suggest that such a simplistic interpretation may not be warranted. In fact, one could argue that it is most consistent with a stochastic model involving the generation of abnormal proteins. This interpretation, however, must also be softened because the nature of cell injury induced by amino acid analogues is poorly understood, and may be the result of mechanisms other than via incorporation into cellular proteins (41-44). Also, our observations indicate that the phenotype of the analogue treated cells is apparently different from that of the senescent cells. As suggested above, it is possible that the heterokaryon analysis does not discriminate between various types of cell injury. To test this possibility we are exposing young human cells to other agents (Mytomycin C, for example) which will induce a sustained depression in the [^{3}H] thymidine labeling index and in which the mechanism of injury is known. We will then study the pattern of nuclear thymidine incorporation in heterokaryons between these cells and actively dividing cells. The interpretation of a unique response would, of course, depend on the quality and quantity of the difference. If, on the other hand, the results of these experiments were similar to those observed with senescent and amino acid treated cells, then one could conclude that a fundamental difference between neoplastoid and hyperplastoid cells has been revealed. For example, these results could reflect a difference in the efficiency of various error correcting systems. Such an interpretation would provide a rational basis for a comparative study of the known scavenger and repair systems not only between hyperplastoid and neoplastoid cells, but also between cells derived from different species. This suggestion is further supported by two previously published observations: 1) eukaryotic cells selectively degrade abnormal proteins (14,15) and 2) the efficiency of UV repair in cultured mammalian cells is directly related to the longevity of the species from which they are derived (46).

The final interpretation of the effects of enucleate cytoplasts must await completion of control experiments. However, if the results indicate that the diminished capacity of enucleate cytoplasms of neoplastoid cells is biologically significant (and not the result of experimental machinations) then several interpretations can be envisioned. The nucleus may produce labile, effector molecule(s) which are produced at specific point(s) in the cell cycle. If this is the case, then enucleate cytoplasms obtained from synchronized populations in various stages of the cell cycle should have different stimulatory capacities. Alternatively, the nucleus may be necessary for the maintenance of an internal milieu which involves many factors and must be maintained throughout the cell cycle. Both models can account for the modest stimulation of senescent nuclei observed in the experiments above. On the other hand, if it is determined that HeLa cytoplasms (or from other neoplastoid cells) are capable of stimulating thymidine incorporation, then one could investigate the effect of organelles which are accessible for study, such as isolated membrane ghosts.

We have attempted fusion of "isolated" nuclei which are present in a pellet at the bottom of the Corex tube. However, these preparations have proven unsatisfactory because of contamination by cytoplasmic droplets. Recently, we have employed with minor modification, a recently reported purification technique (47) with very satisfactory results. Clearly these experiments will be important for the interpretation of the studies with enucleate cytoplasms.

It would be of interest to repeat these experiments using SV-40 transformed human cells because of the different kinetics of labeling in the heterokaryon experiments described above (Figure IV). However, to date, attempts to enucleate these cells efficiently have been unsuccessful because of poor anchorage of the cells to the substrate. However, modification of culture and enucleation techniques show promise that these methodologic problems can be solved.

Wright and Hayflick have thus far published the only studies using this technology in the study of clonal senescence (48,49). They concluded that the expression of the senescent phenotype is a nuclear function.

As with the enucleation studies, the synkaryon experiments are still very preliminary. The failure of selectively proliferating cells to complement those derived from a patient with Werner's syndrome suggest dominance of the senescent phenotype with respect to young diploid cells. This conclusion, however, must be guarded, since it is based on a negative result, and cells from donors with donors with Werner's syndrome may be a special case.

The apparent stability of the pure hybrid clones is of relevance to somatic segregation studies. Our laboratory has presented evidence for diploidization within spontaneously occurring tetraploid clones from euploid human cell cultures heterozygous for certain marker chromosomes (50); a few diploid cells were observed, suggesting somatic segregation. In the present studies the three hybrid clones remained chromosomally stable; the mixoploid clones almost certainly resulted from interclonal contamination. This indicates that diploidization and hence somatic segregation via this mechanism may be a rare event.

One of the severest criticisms of these studies concerns the nature of the DNA synthesis observed in the heterokaryon and heteroplasmosis studies. We have no direct evidence that it represents scheduled DNA synthesis. We are planning to repeat these studies in the presence of inhibitors of scheduled DNA synthesis, such as hydroxyurea. If nuclear thymidine incorporation is inhibited under these conditions, then it would be unlikely that we are observing a repair type synthesis. We cannot, of course, using purely cytological techniques, exclude some form of re-utilization of the labeled thymidine. It is therefore imperative that the heterokaryon experiments be closely coupled with the synkaryon studies so that the observed DNA synthetic patterns can be correlated with the presence or absence of sequential cell cycle function.

Other criticisms include lack of precise information about thymidine pool sizes in the various cell types and in senescent cells, as well as possible alterations in thymidine metabolism associated with senescence or transformation. In regard to the latter, Cristofalo has made observations suggesting that alteration of thymidine kinase activity is not causally related to alterations of proliferative activity in senescent cells (51). Clearly such studies are relevant to the interpretation of our experiments and must be completed.

These studies and those cited in the literature indicate that a pattern of complementation between the senescent human diploid cell and actively proliferating cell types is emerging. However, it is clear that the basic mechanisms of clonal senescence will not be elucidated by this approach alone. We believe that these hybridization studies will help to establish a rational basis for future studies of clonal senescence at the molecular level of analysis.

REFERENCES

1. Hayflick, L. and Moorhead, P.S. Exp. Cell Res. 25: 585 (1961).
2. Kohn, R.R. Science 188: 203 (1975).
3. Hayflick, L. Exp. Cell Res. 37: 614 (1965).
4. Goldstein, S., Littlefield, J.W. and Soeldner, J.S. Proc. Nat. Acad. Sci. USA 64: 155 (1969).
5. Martin, G.M., Sprague, C.A. and Epstein, C.J. Lab. Invest. 23: 86 (1970).
6. LeGuilly, V., Simon, C., Lenoir, P. and Gourd, M. Gerontologia 19: 303 (1973).
7. Martin, G.M. and Sprague, C.A. Exp. Mol. Path. 18: 125 (1973).
8. Martin, G.M., Ogburn, C. and Sprague, C. Adv. Exp. Med. & Biol. 61: (1975).
9. Goldstein, S. Lancet I: 424 (1969).
10. Hayflick, L. Bioscience 25: 629 (1975).
11. Cristofalo, F.J. Adv. Gerontol. Res. 4: 45 (1972).
12. Orgel, L.E. Proc. Nat. Acad. Sci. USA 49: 517 (1963).
13. Goldberg, A.L. Proc. Nat. Acad. Sci. USA 69: 422 (1974).
14. Capecchi, M.R., Capecchi, N.E., Hughes, S.H. and Wahl, G.M. Proc. Nat. Acad. Sci. 71: 4732 (1974).
15. Burnet, F.M. Pathology 6: 1 (1974).
16. Burnet, F.M. In: Intrinsic Mutagenesis, A Genetic Approach to Aging. John Wiley & Sons, New York. (1974).
17. Harmon, D. J. Geront. 11: 298 (1956).
18. Harmon, D. J. Geront. 16: 247 (1956).
19. Martin, G.M., Sprague, C.A., Norwood, T.H. and Pendergrass, W.R. Amer. J. Path. 74: 137 (1974).
20. Martin, G.M., Sprague, C.A., Norwood, T.H., Pendergrass, W.R., Bornstein, P., Hoehn, H. and Arend, W.P. Adv. Exp. Med. Biol. 53: 67 (1975).
21. Merz, G.S., Jr. and Ross, J.D. J. Cell Physiol. 74: 219 (1969).
22. Wright, W.E. and Hayflick, L. Adv. Exp. Med. Biol. 61: 39 (1975).
23. Harris, H., Watkins, J.F., Ford, C.E. and Schoefl, G.I. J. Cell Sci. 1: 1 (1966).
24. Harris, H. J. Cell Sci. 2: 23 (1967).
25. Norwood, T.H., Pendergrass, W.R., Sprague, C.A. and Martin, G. M. Proc. Natl Acad. Sci. USA 71: 2231 (1974).
26. Baserga, R. and Nemeroff, K. J. Histoch. & Cytochem. 10: 628 (1962).
27. Harris, H., Sidebottom, D., Grace, D.M. and Bramwell, M.E. J. Cell Sci. 4: 499 (1969).
28. Norwood, T.H., Pendergrass, W.R. and Martin, G.M. J. Cell Biol. 64: 551 (1975).
29. Prescott, D.M. and Kirkpatrick, J.B. In: Methods in Cell Biology. (Ed.) D.M. Prescott. Academic Press, New York (1973) Vol. 7 p. 189.
30. Wright, W.E. In: Methods in Cell Biology (Ed.) D.M. Prescott. Academic Press, New York (1973) Vol 7 p. 203.

31. Croce, C.M. and Koprowski. H. Virology 51: 227 (1973).
32. Pontecorvo, G. Somatic Cell Genet. 1: 397 (1975).
33. Davidson, R.L. and Gerald, P.S. Somatic Cell Genet. in press (1976).
34. Norwood, T.H., Zeigler, C.J. and Martin, G.M. Somatic Cell Genet. in press (1976).
35. Littlefield, J. J. Cell Physiol. 82: 129 (1973).
36. Croce, C.M. and Koprowski, H. Science 184: 1288 (1974).
37. Goldstein, S. and Lin, C.C. Exp. Cell Res. 70: 436 (1971).
38. Hoehn, H., Bryant, E.M., Johnston, P., Norwood, T.H. and Martin, G.M. Nature 258: 608 (1975).
39. Sprague, C.A., Hoehn, H. and Martin, G.M. J. Cell Biol. 60: 781 (1974).
40. Smith, J.R. and Hayflick, L. J. Cell Biol. 62: 48 (1974).
41. Vaughn, M. and Steinberg, D. In: Advances in Protein Chemistry (Eds.) C.B. Anfinsen Jr., K. Baily, M.N. Anson and J.T. Edsall. Academic Press, New York (1959) Vol. 14 p. 115.
42. Cohen, G.N. and Gros, F. In: Annual Reviews of Biochemistry (Ed.) J.M. Luck. Annual Reviews Inc., Palo Alto (1960) Vol. 29 p. 525.
43. Shive, W. and Skinner, C.G. In: Metabolic Inhibitors, A Comprehensive Treatise (Eds.) R.M. Hochster and J.H. Ouastler Academic Press, New York (1963) Vol. 1 p. 2.
44. Meister, A. Biochem. Amino Acids 1: 231 (1965).
45. Hendil, K.B. J. Cell Physiol. 87: 289 (1976).
46. Hart, R.W. and Setlow, R.B. Proc. Nat. Acad. Sci. USA 71: 2169 (1974).
47. Lucas, J.J., Szekely, E. and Kates, J.K. Cell 7: 115 (1976).
48. Wright, W.E. and Hayflick, L. Fed. Proc. 34: 79 (1975).
49. Wright, W.E. and Hayflick, L. Exp. Cell Res. 96: 113 (1975).
50. Martin, G.M. and Sprague, C.A. Science 166: 761 (1969).
51. Cristofalo, V.J. In: Explorations in Aging (Eds.) V.J. Cristofalo, J. Roberts and R.C. Adelman. Plenum Press, New York (1974) p. 57.

COMMENTS ABOUT THE PAPER BY NORWOOD ET AL. AND AN IDEA RELATING SOMATIC CELL MUTATIONS TO CELLULAR SENESCENCE

Robert DeMars

Laboratory of Genetics
University of Wisconsin
Madison, Wisconsin 53706

Cells, nuclei and cytoplasms from young, senescent and permanent line populations of human cells can be usefully combined as pairs in 30 different ways. Dr. Norwood, with his collaborators at Seattle, and others, have been attempting to create these combinations in order to answer three main questions: (i) Is the senescence phenotype dominant?; (ii) Can the senescence phenotype be imposed and maintained by the cytoplasm of a senescent cell or must the nucleus be present, too?; (iii) Is the senescence phenotype determined by primary, stable nuclear changes? Some of the possible combinations of cells and parts of cells have not been studied yet and the interpretation of the observations made so far must be tentative. Therefore, my main comments about Dr. Norwood's presentation are an expression of admiration for what has been accomplished and a hope that the good work will continue.

Ambiguities that remain are exemplified by the study of dominance. Senescence, defined as cessation of DNA replication, was trans-dominant in heterokaryons of young and senescent fibroblasts and was recessive in heterokaryons made with SV-40 transformed fibroblasts. The interpretation of these observations is complicated by the behavior of heterokaryons made with HeLa cells. Replication of senescent cell nuclei occurred in most heterokaryons 24 hours after fusion but was followed by the failure of both senescent cell nuclei and HeLa nuclei to replicate in about 90 per cent of heterokaryons 72 hours after fusion. Part of the stimulatory effect on DNA synthesis in senescent nuclei could be attributed to HeLa cytoplasm. Heteroplasmons prepared with HeLa cytoplasts and senescent cells had an increased incidence of DNA-synthesizing senescent nuclei but not as great an incidence as in heterokaryons. These observations indicate that senescence can be transiently recessive and, perhaps, transiently dominant. The degrees of dominance and transience can be influenced by the presence of a nucleus from a transformed or senescent cell.

The observations seem to show that absence of DNA synthesis in senescent fibroblasts does not mean permanent loss of synthesizing ability in many, perhaps most or all, senescent nuclei. However, the Seattle group emphasizes that they must still determine if the DNA synthesis they detect by autoradiography is scheduled, replicative synthesis rather than unscheduled, repair synthesis.

The possible transience of dominance effects in heteroplasmons is important for the interpretation of observations reported by Wright and Hayflick (1). They found that the doubling number of heteroplasmons prepared with young and senescent fibroblasts corresponded to that of the nucleated parent. The result obtained with young cytoplast-senescent cell heteroplasmons can be interpreted in at least two ways. One way requires the assumption that proliferation of heteroplasmons is as efficient on a per cell basis as the replication of senescent cells alone. Therefore, population doublings accurately reflect typical doubling numbers of cells in the population. I question the accuracy of this interpretation since no currently used method of cell fusion results in proliferation of anywhere near all of the fused cells. It seems more likely that the procedures used for treating the cells prior to fusing them and the fusion procedure, itself, reduced the ability of most cells to proliferate. This means that small population doubling numbers could reflect large doubling numbers of a minority of the cells in a heteroplasmon population. For instance, the fact that populations of young cytoplasm - old cell heteroplasmons went through up to five doublings could mean that the initial population doubling might have resulted from seven doublings of about one per cent of the heteroplasmons. The real total number of doublings would have been 11 and it would have been more difficult to ignore possible effects of young cytoplasm on the replication of senescent cells. Wright and Hayflick (1) discussed this interpretation (and others) critically but their decision to discard it would have received strong support if they had counted cells in clones instead of determining population doublings alone.

Wright and Hayflick (1) tentatively concluded that senescence is determined by the state of the nucleus and referred to it as a programmed genetic phenomenon. Norwood and his collaborators tentatively favor a broadly similar interpretation of senescence as a terminal differentiation. Neither group emphasized a role for stochastic processes as primary causes of senescence. They differ from those who propose that low probability errors in molecules concerned with macromolecular synthesis initiate subsequent error catastrophes that result in cell senescence (e.g., 2). There is strong evidence for error laden cytoplasms of senescent cells (summarized in [2]) but it is still questionable if senescence is initiated by epigenetic events of the sorts proposed by error catastrophists.The problem is one of explaining a regular phenomenon (senescence) in terms of improbable, random events. Perhaps this can be done using a mechanism that is often regarded as unlikely to provide a satisfactory solution.

Mutations in nuclear genes are sometimes thought to occur too rarely to be a prominent kind of senescence-initiating event in cultured human fibroblasts. This outlook may be wrong and should be examined with reference to information about rates of spontaneous mutation of human somatic cells _in vitro_. Available estimates form two broad categories: "HIGH", 10^{-6} - 10^{-5} per cell generation, and "LOW", less than 10^{-7} per cell generation. It is likely that estimates will form a more nearly continuous distribution as information accumulates. The HIGH range is based mostly on fluctuation tests performed in my laboratory and applies to the rates of change to azaguanine-resistance (=AG-resistance) (3) and to 2,6-diamino-purine resistance (=DAP-resistance) (4). Most of the evidence for the mutant nature of resistant variants is based on the kind of analysis described in pp. 351-359 (5). All of our evidence with human fibroblasts indicates that AG-resistance results from single mutations that occur on the X-chromosome and reduce the activity of the HPRT enzyme coded for by the _hpt_ locus. All nine fluctuation tests we have performed with six unrelated strains of fibroblasts yielded mutation rate estimates in the HIGH range (3 and unpublished).

DAP-resistance results from mutations that reduce the activity of APRT, which is coded for by the _apt_ locus on autosome No. 16. Spontaneously originating DAP-resistant mutants are too rare to be detectable in a practical way in the majority of human cell strains because DAP-resistance is recessive under the conditions needed for quantitative selection. Mutants are readily detectable in fibroblast strains from that minority of humans who are heterozygous at the _apt_ locus. My associates, Dr. Jay Jackson and Ms. Carolyn Steglich, and I have recently performed a double fluctuation test with one heterozygous strain and obtained rates of 8.30×10^{-6} per cell generation to DAP-resistance and 1.97×10^{-6} per cell generation to AG-resistance. The difference between the rates is statistically significant but the mere four-fold difference is consistent with the notion that DAP-resistance resulted from single mutational events.

The LOW range of rates concerns resistance to α-aminitin resulting from mutations affecting RNA polymerase II (6) and resistance to ouabain resulting from mutations that affect the

(Na^+ - K^+) ATP-ase (7). Neither of the rates of spontaneous mutation has been directly determined with human cells but their low values are inferred from the relative rarity of mutant cells in comparison to AG-resistant and DAP-resistant mutants. Resistance to α-aminitin is partially dominant, as is resistance to ouabain, and mutants probably resulted from single mutational events. The rate of spontaneous mutation to ouabain-resistance in Chinese hamster cells and mouse L cells has been estimated to about about 5×10^{-8} per cell generation (8).

The difference between the HIGH and LOW ranges probably doesn't result from intrinsically different rates of mutation in DNA. (Na^+ - K^+) ATP-ase and RNA polymerase II are required for cell viability. Therefore, only that minority of all mutations in a gene that render the enzyme resistant to an inhibitor but capable of adequate normal function can be detected as viable mutants. In contrast, HPRT and APRT are completely dispensable under the conditions used to isolate mutants: all mutations that reduce the enzyme activity enough to confer resistance to the selective purine analogue can be detected as viable mutants. This interpretation of the HIGH and LOW mutation rates is supported by the findings that X-rays, ultraviolet light and all general categories of chemical mutagens increase the rate of change to AG-resistance but X-rays do not increase the incidence of ouabain-resistant mutants (9, 10). The rates of mutation to AG-resistance and DAP-resistance may represent the high part of the range of detectable single locus mutation rates for the purpose of thinking about mutations and cell senescence.

Even HIGH single locus mutation rates are, alone, too low to account for the eventual inability of all cells in a fibroblast population to proliferate after 50-150 doublings in vitro. Improbable mutations and inevitable senescence could be reconciled if a mutation in any one of a large number of loci can initiate cell senescence so that the total rate of senescence-engendering mutations is the sum of many individually low rates. We can't realistically estimate the number of loci that might be involved but can base one kind of estimate on the general theory of error catastrophe (2). The main tenet of the theory is that senescence is initiated by errors in basic processes of macromolecular synthesis in cells, especially the synthesis of proteins. Once such errors occur they can decrease the accuracy of other aspects of information transfer. The result is an ever-increasing, finally lethal error rate. Could the initiating errors be mutations in genes directly concerned with information transfer, e.g., replication, transcription and translation? Genes for DNA polymerases, RNA polymerases, ribosomal proteins, ribosomal RNAs, transfer RNAs, aminoacyl-tRNA synthetases and some others could account for about 100 kinds of genes. If mutant forms of histones or other relatively common chromosomal proteins can cause cellular senescence by altering gene expression or reducing its accuracy perhaps 100 more genes can be added to the estimate.

Some kinds of genes mentioned above are iterated and a mutation in any one of the copies might increase the error rate enough to start a cascade of errors in later cell generations. For instance, a diploid human cell may have 600 copies of ribosomal RNA genes, 4000 copies of 5S RNA genes, 2600 4S RNA genes and 20-40 copies of histone genes (11). Summing, the number of target genes in which hypothetical senescence initiating mutations might occur may be 10^3-10^4. I can't guess what fraction of all mutations in these genes would affect fidelity of information transfer but the summed rate of change could be as high as 10^{-3} - 10^{-1} per cell generation if we assume that the HIGH range of single locus mutations most accurately reflects the amount of mutation within a gene. If the average total rate of potentially error-increasing mutations is 10^{-2} per cell generation a large proportion of all cells will have one or more such mutations after 100 population doublings.

The kinds of mutations postulated above could account for the observed delay between senescence-initiating events and the actual cessation of cell proliferation: the frequent occurrence of clones consisting of several hundred senescent cells indicates an interval of at least 5-10 doublings (unpublished observations) but, the delay may often be greater (11). A mutation in a single copy of an iterated gene would have little or no immediate effect on a cell because the numerous non-mutant copies could still produce normal products. The genetic buffering against the deleterious effects of mutations prompted Medvedev (13) and Cutler (14) to argue that single mutations in multiple-copy genes concerned with macromolecular synthesis are not likely to cause senescence. The main effects of such mutations were assumed to be slight, tolerable reductions in essential functions. I emphasize here that another important effect of such mutations may be the initiation of clones in all the cells of which abnormal gene products (e.g., an ambiguously functioning tRNA) can increase the error rate of macromolecular synthesis. A single phenomenon, gene repetition, improves the chances that cells can survive the immediate, deleterious effects of certain mutations but may also be an ultimate cause of their senescence because it increases the overall rate of mutation that can initiate error catastrophes. The apparently programmed aspect of cell senescence could be mainly a function of the product of the mutation rate per gene and the degree of repetition of genes concerned with senescence.

The notion that mutations occurring at ordinary rates in iterated genes often initiate cellular senescence is half-baked as presented here. It is offered in order to redirect attention to a frequently discarded hypothesis concerning senescence (2, 13 and 15 contain references to and discussions of various somatic mutation theories). The reason most often given for doubting the correctness of the somatic mutation hypothesis is that it requires an "unreasonably" high mutation rate. The same reason is often used for thinking that somatic mutations contribute little to the generation of antibody variability. Yet, somatic mutation can account for a large part of antibody variability if one uses the HIGH range of single locus mutation rates described above as an estimator of the amount of amino acid change in non-essential proteins, such as immunoglobulins. Consider the following argument:

Variation in antibody specificity is the result of amino acid substitutions in the combining site regions of light chains and heavy chains and of the different combinations of light and heavy chains. A rough estimate of the rate of amino acid substitution can be based on the rate of mutation to azaguanine resistance. The molecular weight of the HPRT monomer is about 2.8×10^4 i.e., there are about 250 amino acids per monomer. Since the average rate of mutation to azaguanine resistance is about 4×10^{-6} per cell generation the average rate of change per amino acid is about 1.6×10^{-8} per cell generation. Overestimation results from mutations, such as large deletions, that do not cause single amino acid substitutions. Underestimation results from amino acid substitutions that do not reduce HPRT activity enough to be manifested as resistance to azaguanine. Assuming 50 amino acid positions distributed between the light and heavy chains define a specific combining site, the rate of mutation per combining site is about 8×10^{-7} per cell generation. There may be about thirty combining site specificities encoded in a gamete. They are defined by at least four related genes that code for κ subtypes, at least four that code for λ subtypes and at least four that code for α, γ, μ and ε heavy chains. Each inherited combining site can be regarded as a target for somatic mutations that will define new combining sites. Therefore, the total mutation rate in combining sites is about 2.4×10^{-5} per cell generation. If only 10^9 cell divisions are needed to establish the "bank" of specific antibody clones then about 2.4×10^4 different clones can be generated by somatic mutation. Use of the HIGH rate of mutation is justified in making this estimate because antibody production is not a function necessary for the life of cells, not even cells that are differentiated for that function. Many new specificities can be added during proliferation of memory cells during life. The argument just presented suggests that different gametes of an individual might be very diverse in their codes for antibody specificity and that humans would be found to be highly polymorphic if the predominant specificities in their germ cells could be determined.

These notions have an interesting possible corollary with regard to cell senescence. The maximum life spans of mammals differ by a factor of about 50 (16) and the life spans of their somatic cells in vitro vary in a roughly parallel manner (17). Therefore, the antibody-forming cells of short-lived species, such as mice, may have fewer doublings available for generating necessary antibody variability than do the corresponding cells of longer-lived humans, for instance. If the required numbers of specificities are similar in the two species part of the necessary overall mutation rate affecting antibody specificity could be achieved by the presence of a larger number of antibody genes in mice than in humans.

A reason that one can partly explain antibody variability with reference to somatic mutation is the evidence for iteration of antibody genes, i.e., the existence of different sub-families of light chains and heavy chains. The realization that somatic mutation at ordinary rates combined with gene repetition can account for much antibody variability suggests that a similar explanation of fibroblast senescence merits further consideration. I will not attempt here to generalize the crude somatic mutation hypothesis presented above or to reconcile it with all key observations and literature concerning senescence. Rather, I want to emphasize the thought that if error catastrophes triggered by error-prone molecules engaged in macromolecular synthesis are responsible for fibroblast senescence then it is reasonable to suspect that mutations in iterated genes are sometimes primary initiators of senescence. Error-generating molecules that originate after transcription have limited chances to trigger catastrophes because it is likely that they are produced singly or in small numbers in any one cell. Such molecules are probably degraded along with their normal protein and nucleic acid counterparts in most instances. In contrast, mutant genes that code for error-generating products can cause them to recurrently appear in multiple copies. The chance for abnormal gene products to trigger error catastrophes is perpetuated in succeeding cell generations because the mutations are unlikely to have immediate lethal effects. Here are some thoughts and experiments that might be useful in determining if the form of somatic mutation hypothesis presented above is useful.

One vulnerable assumption of the hypothesis is the existence of error prone protein synthesis in senescent cells. Evidence often cited as contradicting this assumption is the unreduced yields of viruses produced in senescent human fibroblasts and unelevated incidences of thermolabile particles in those yields (18-20). Three assumptions of those studies are: (a) When normal and defective forms of proteins used in vegetative functions of the viruses are present in the same cell the amount of normal protein is not adequate for the production of a normal yield; (b) Defective structural proteins are incorporated into virions along with normal proteins; (c) The infectivity of virions containing defective structural proteins is abnormally thermolabile. None of these assumptions was supported by independent evidence. For instance, no attempt was made to show that the viruses liberated from cells mixedly infected with normal and thermolabile mutants contained genotypically normal viruses that were abnormally thermolabile. This is a key control because it can account for the possibility that the viruses used in the cited studies exercise "quality control" in self-assembly i.e., that they preferentially use normal proteins when both normal and defective proteins are available. The use of viral probes of the protein error hypothesis of cellular senescence is a good idea but the evidence published so far is inconclusive.

Base-changer mutagens should reduce the doubling number of human fibroblasts in vitro because the mutant genes would produce abnormal gene products that could engender error catastrophes. General alkylating agents, such as MNNG wouldn't be simply suitable for investigating the role of induced DNA mutations, since they also readily alkylate cytoplasmic RNA and, even, proteins. Perhaps BUDR should be used because of its preferential incorporation into DNA; it is known to be mutagenic in human fibroblasts (21). X-rays also induce mutations in human cells (22, 23) but might reduce the doubling number less than base-changer mutagens because many mutations induced by X-rays are deletions: the mutant genes often would not yield abnormal products that could increase the error rate of macromolecular synthesis. One could compare BUDR and X-rays with regard to the reduction in doubling number in relation to the induction of AG-resistant mutants.

The somatic mutation hypothesis requires that cells have senescence-initiating mutations in nuclear genes long before senescence occurs. More than 50 per cent of human fibroblasts are still able to synthesize DNA about ten doublings before all the cells in the population stop proliferating (24). The numbers of cells in senescent clones also suggest that senescence-initiating events often occur at least 5-10 doublings before senescence occurs. The nuclei, but not the cytoplasms, of proliferating cells about 10 doublings before senescence should most markedly reduce the doubling number of much younger cells with which they are combined. It should be possible to isolate large enough numbers of hybrids for these experiments by using young $HPRT^-$ cells and pre-senescent $HPRT^+$ nuclear and cytoplasmic donors and selecting for hybrids in medium containing azaserine and hypoxanthine.

It might be possible to usefully combine the mutagenic and cell fusion approaches. I wrote above that a possible overall mutation rate in genes directly involved in macromolecular synthesis is 10^{-2} per cell generation. It is now possible to increase the mutation rate for the hpt and apt loci about 100-fold using MNNG (25). The treatments used to attain a 100-fold increase in the mutation rate of a single locus might induce an average of one senescence-initiating mutation per cell in iterated genes. Nuclei from very young cells so treated should greatly reduce the doubling number of young recipient cells. The numbers of senescent cells in senescent clones produced in this way would permit direct determination of the numbers of doublings that must intervene between individual senescence-initiating mutations and their final expression as senescence of a clone.

The kinds of hypothetical senescence initiating mutations I have proposed have both dominant and recessive properties. The initiating mutation in one copy of an interated gene is "recessive" because senescence is not manifested until the increase in error rate or accumulated gene products is large enough to stop cell proliferation. The mutations are "dominant" because, once present, they cause senescence in the long run even though normal copies of the genes may be present. These properties are wholly in accord with the reports that tetraploid hybrids of human fibroblasts senesce at about the same time as their diploid parents (26). In fact, the "dominant" nature of senescence-initiating mutations and the increased number of targets for such mutations in tetraploid cells might sometimes cause tetraploids to senesce earlier than diploids.

According to the hypothesis presented here, different senescent clones of the same fibroblast population may have senescence initiating mutations in different kinds of genes. Independent mutations may have different effects on error rates and, therefore, engender senescence after different intervals and in different ways. A special way of interpreting the hypothesis that defective proteins of senescent cells result from frequent "misreading" during information transfer is that some senescent clones might act as if they have suppressor mutations. Suppressor mutants have not been described in mammalian cells and it may be because they regularly initiate senescence. However, it might be possible to investigate this aspect of the somatic mutation hypothesis by comparing the abilities of a variety of missense and nonsense mutants of a virus to multiply in senescent and pre-senescent fibroblasts. There should be specific complementary relations between some mutants and some senescent clones.

A second derivative of the idea that senescence can be initiated by mutations in different kinds of genes, some of them iterated, is that there might be at least two broad categories of ways in which strains with spanned in vitro lifetimes can become permanent cell lines. One broad category could consist of changes that make the cell cycle time for some cells shorter than the time needed for the occurrence of the final error that stops cell division (e.g., 15). As long as the inocula used in subculturing a population always include some cells in which this rate relation exists the population will have unspanned growth. Several, perhaps many, different kinds of mutations could result in the necessary adjustment of cell division and error rates. The fact that spontaneous origin of unspanned growth of human fibroblasts hasn't been reported leads to the suspicion that the hypothetical rate adjustment alone is seldom or never responsible for the origin of permanent lines.

A second broad category of strain-to-line changes could include changes that compensate only for a specific kind of senescence initiating mutation. Experiments and the existence of certain phenomena are suggested by this idea.

(a) Populations that have escaped their first senescence crisis may acquire new kinds of senescence initiating mutations that are initially uncompensated and, subsequently cause senescence in at least some cells of a permanent line. Such episodes might not be reported because they might be regarded as unexplained nuisances that become uninteresting when the cultures survive the crises after acquiring second senescence compensating changes. It should be noted that cloning efficiencies of permanent lines are often less than 1.0. It is easy to explain away the death of cells of permanent lines as resulting from technical mistreatment but it may result from inadequate compensation of senescence initiating mutations in some cells, even a majority of them.

(b) It might be possible to initiate senescence in many cells of a permanent line by introducing chromosomes that have senescence initiating mutations that are not compensated for in the line. The literature about hybrids formed by fusing human fibroblasts with permanent cell lines describes the properties of the hybrids that proliferate without limit; little or no attention is devoted to hybrid populations that stop growing for no apparent reason, sometimes after many doublings. Is it possible that some hybrids retain human chromosomes with senescence initiating mutations in many cells? If the human chromosomes with these mutations lack other genes necessary for the life of the hybrid cells then those cells from which the chromosomes are lost can survive the senescence crisis of the cell population. Elaborations of this experimental approach using independent senescent clones and pre-senescent as well as senescent human fibroblasts to prepare hybrids are obvious.

(c) If the preceding notions are correct it might be possible to avert senescence in some cells of a strain by introducing specific chromosomes bearing the senescence compensating mutation(s) of a permanent cell line. This could best be attempted using metaphase chromosomes (27-29) or microcells (30-32). If senescence compensating mutations are specifically complementary to certain senescence initiating mutations different cell lines of the same species might donate different chromosomes to produce evergrowing derivatives of a cell strain. This does not necessarily contradict the reports by Croce (this volume) that permanent lines derived from the fusion of normal mouse macrophages and SV-40-transformed human fibroblasts specifically retain the human chromosome bearing the SV-40 genome. Such lines may be only those among all possible lines that have a senescence-initiating change that is specifically compensated for by an SV-40 expression. Therefore, a given cell line, whether or not it is transformed by a virus, might be able to donate an effective senescence compensating chromosome to some senescent clones and not others.

Further elaboration of experimental possibilities would be superfluous here, especially because at least two very important aspects of any general hypothesis for explaining cell senescence have been neglected so far: the "permanent line" property of germ cells and the different timing of senescence in corresponding cell types from mammalian species that have different life spans. No thorough attempt to explain these phenomena will be made here but some comments might be useful.

According to the main hypothesis presented in this paper two of the most important determinants of the timing of cell senescence are the rates of mutation in genes concerned with information transfer and the degree of repetition of those genes. The total <u>effective</u> senescence-initiating mutation rate may also be affected by the fractions of all iterated genes that are actually transcribed, since they could vary among species and cell types (14). Germ cell mutation rates are usually expressed as mutations per locus per human generation but somatic cell mutation rates are expressed as mutations per locus per cell generation. Comparison of the two kinds of rates requires determination of the (large) number of cell generations that precedes the formation of functional gametes, determination of the incidences of somatic cells and germ cells with mutations in the same gene and determination of the fraction of all germ cells and their precursors that are non-functional, perhaps because they are senescent or because they bear other mutations that lead to their elimination as gametes or early zygotes. The essential characteristic of a permanent line is not that all cells of the population are viable but, simply, that "inocula" used for propagation, or procreation in the case of germ cells, contain enough viable cells to regenerate the population. Zygotes that finally get born may be formed from highly selected gametes that have fewer mutations than the average gamete or precursor. In the absence of critical information of the sorts just mentioned failure to explain the apparent permanence of the germ line at this time should not, by itself, disqualify otherwise useful hypotheses about cell senescence.

Shortage of key information also thwarts the interpretation of different senescence times of fibroblasts from humans and rodents, for instance. Few critical comparisons of rates of mutation, especially spontaneous mutation, can be made now. Almost all work with rodent cells has been done with permanent cell lines rather than pre-senescent populations. Conditions for quantifying the same kinds of mutations in different kinds of cells have not been rigorously defined. It seems unlikely now that mouse cells senesce so much earlier than human cells because their mutation rate is 50 times greater, but I doubt that available evidence could rigorously exclude even a 10-fold difference. Mouse cells are greatly deficient relative to human cells in the capacity for certain types of accurate repair of DNA damage (reviewed in [33]) and this difference might be manifested as a difference in mutation rate.

The somatic mutation hypothesis of cell senescence evokes thoughts about carcinogenesis. There is an intriguing inverse relationship between rates of senescence and rates of transformation into permanent lines and toward malignancy. Cells of mice and hamsters senesce relatively early but readily yield permanent lines in which tumorigenic variants arise spontaneously and after treatment with mutagens and carcinogens. There is much evidence that transformation toward malignancy requires at least two changes and that permanent line character is separable from tumorigenicity. If these traits are acquired by means of independent, individually rare events the chance of demonstrating full transformation toward malignancy of propagable cells *in vitro* should be limited by the rates of the events and the number of doublings available for their occurrence. If permanent line character is acquired first there is unlimited opportunity for the later occurrence of changes that confer the ability to form a tumor. Tumorigenicity traits may be acquired first but may not be fully demonstrable if clones possessing the traits senesce. If such abortive transformed clones of rodent cells occur they have received little attention. Most studies of *in vitro* transformation of pre-senescent cells are performed with cells of the Syrian hamster. Transformation toward malignancy is scored as morphological change and loss of contact inhibition in colonies fixed and stained within 14 days after carcinogenic treatment. Living cells are isolated only from a sample of unfixed colonies that have transformed traits and there would be a natural bias toward isolating cells from the largest of these colonies i.e., from those least likely to have undergone senescence.

The readiness with which transformation toward malignancy is observed in rodent cells may reflect their relative ease of transformation into permanent lines. The rates of spontaneous origin of permanent cell lines have not been determined as mutation rates are determined but they must be low on the basis of cell generations. Effects of mutagens and carcinogens on rates of strain-to-line transformation have not been directly measured, either, but the incidences of non-contact-inhibited transformants per cell surviving treatments with chemical carcinogens can be 0.01 - 0.10 when the survival is 0.1 or greater (e.g., 34-36), much higher than induced rates for single loci that specify known proteins. In contrast, normal human diploid fibroblasts senesce much later than rodent cells but, so far, there is no verified report of their spontaneous transformation into a permanent line or of their chemically induced complete transformation into tumorigenic cells _in vitro_. We have failed to isolate fully transformed human fibroblasts after treatment with MNNG that increased the incidence of AG-resistant mutants to about 10^{-3} per survivor. However, among the survivors were clones of senescible cells that had a high cloning efficiency in medium made semi-solid with soft agar or methyl cellulose and that formed small lumps of living human cells under the skin of nude mice (37). Perhaps such clones are examples of the possible abortive transformed clones mentioned above, which must survive their senescence crisis in order to be clearly demonstrable as tumorigenic. It is possible, then, that relatively high effective rates of mutation in iterated genes of mouse cells, for instance, are responsible for both their early senescence and for a relatively large probability that senescence-compensating changes will occur in inocula of sizes commonly used in subculturing pre-senescent cultures. A lower rate of senescence-compensating changes in human fibroblasts may be the main impediment to facile experimentation on the _in vitro_ transformation of human cells toward malignancy. One of the important suggestions of the hypothesis presented in this paper is that it might be possible to create permanent human cell lines from normal cells at will by introducing human chromosomes bearing specific senescence-compensating mutations. Demonstration of the absence of traits specifically associated with tumorigenicity would then make such cell lines extremely useful subjects for the study of malignant transformation of human cells.

Acknowledgement

This is paper No. 2001 from The Laboratory of Genetics, University of Wisconsin-Madison. Research in this laboratory is supported by grants from The National Institutes of Health (GM 06983, GM 15422 and NIEHS 722095).

References

1. Wright, W.E. and L. Hayflick. The regulation of cellular aging by nuclear events in cultured normal human fibroblasts (WI-38). Adv. Exp. Med. Biol. 61 (1975) 39-55.

2. Orgel, L.E. Ageing of clones of mammalian cells. Nature 243 (1973) 441-445.

3. DeMars, R. and K. R. Held. The spontaneous azaguanine-resistant mutants of diploid human fibroblasts. Humangenetik 16 (1972) 87-110.

4. Rappaport, H. and R. DeMars. Diaminopurine-resistant mutants of cultured diploid human fibroblasts. Genetics 75 (1973) 335-345.

5. DeMars, R. Resistance of cultured human fibroblasts and other cells to purine and pyrimidine analogues in relation to mutagenesis detection. Mutation Res. 24 (1974) 335-364.

6. Buchwald, M. and C.J. Ingles. Human diploid fibroblast mutants with altered RNA polymerase II. Somatic Cell Genet. 2 (1976) 225-233.

7. Mankowitz, R., M. Buchwald and R. M. Baker. Isolation of ouabain-resistant human diploid fibroblasts. Cell 3 (1974) 221-226.

8. Baker, R.M., et al. Ouabain-resistant mutants of mouse and hamster cells in culture. Cell 1 (1974) 9-21.

9. Friedrich, U. and P. Coffino. Mutagenesis in S49 mouse lymphoma cells: induction of resistance to ouabain, 6-thioguanine and dibutyryl cyclic AMP. Proc. Nat. Acad. Sci. 74 (1977) 679-683.

10. Chang, C.C., J.E. Trosko, and T. Akera. Characterization of UV-induced ouabain-resistant mutants in Chinese hamster cells. Submitted for publication in Mutation Research (1977).

11. Tartof, K.D. Redundant genes. Ann. Rev. Genet. 9 (1975) 355-385.

12. Smith, J. R. and L. Hayflick. Variation in the life-span of clones derived from human diploid cell strains. J. Cell Biol. 62 (1974) 48-53.

13. Medvedev, A. Z. Repetition of molecular-genetic information as a possible factor in evolutionary changes of life span. Exp. Gerontol. 7 (1972) 227-238.

14. Cutler, R. G. Redundancy of information content in the genome of mammalian species as a protective mechanism determining aging rate. Mech. Aging and Develop. 2 (1973/ 1974) 381-408.

15. Holliday, R. Growth and death of diploid and transformed human fibroblasts. Federation Proc. 34 (1975) 51-55.

16. Comfort, A. Aging, The Biology of Senescence. (Holt, Rinehart and Winston, New York, Publishers) (1964).

17. Hayflick, L. Biology of human aging. Am. J. Med. Sci. 265 (1973) 433-445.

18. Holland, J.J., D. Kohne, and M. V. Doyle. Analysis of virus replication in ageing human fibroblast cultures. Nature 245 (1973) 316-318.

19. Tompkins, G.A., E. J. Stanbridge and L. Hayflick. Viral probes of aging in the human diploid cell strain Wi-38. Proc. Soc. Exp. Biol. Med. 146 (1974) 385-390.

20. Pitha, J., E. Stark and E. Wimmer. Protein synthesis during aging of human cells in culture. Exptl. Cell Res. 94 (1975) 310-314.

21. Stark, R.M. and J.W. Littlefield. Mutagenic effect of BUdr in diploid human fibroblasts. Mutation Res. 22 (1974) 281-286.

22. Albertini, R.J. and R. DeMars. Somatic cell mutation: detection and quantification of X-ray-induced mutation in cultured, diploid human fibroblasts. Mutation Res. 18 (1973) 199-244.

23. Thacker, J. and R. Cox. Mutation induction and inactivation in mammalian cells exposed to ionising radiation. Nature 258 (1975) 429-431.

24. Cristofalo, V.J. Hydrocortisone as a modulator of cell division and population life span. Adv. Exp. Med. Biol. 61 (1975) 57-79.

25. Jacobs, L. and R. DeMars. Quantification of chemical mutagenesis in diploid human fibroblasts: induction of azaguanine-resistant mutants by N-methyl-N'-nitro-N-nitrosoguanidine. Mutat. Res., in press (1977).

26. Hoehn, H., E.M. Bryant, P. Johnston, T. H. Norwood, and G.M. Martin. Non-selective isolation, stability and longevity of hybrids between normal human somatic cells. Nature 258 (1975) 608-610.

27. McBride, O.W. and H.L. Ozer. Transfer of genetic information by purified metaphase chromosomes. Proc. Nat. Acad. Sci. USA 70 (1973) 1258-1262.

28. Willecki, K. and Ruddle, F.H. Transfer of the human gene for hypoxanthine-guanine phosphoribosyltransferase via isolated human metaphase chromosomes in mouse L-cells. Proc. Nat. Acad. Sci. USA 72 (1975) 1792-1796.

29. Willecke, K., R. Lange, A. Kruger, and T. Reber. Cotransfer of two linked human genes into cultured mouse cells. Proc. Nat. Acad. Sci. USA 73 (1976) 1274-1278.

30. Ege, T. and N.R. Ringertz. Preparation of micro cells by enucleation of micronucleate cells. Exptl. Cell Res. 87 (1974) 378-382.

31. Schor, S.L., R. T. Johnson, and A.M. Mullinger, Perturbation of mammalian cell division. II. Studies on the isolation and characterization of human minisegregant cells. J. Cell Sci. 19 (1975) 281-303.

32. Fournier, R.E.K. and F.H. Ruddle. Microcell-mediated transfer of murine chromosomes into mouse, Chinese hamster, and human somatic cells. Proc. Nat. Acad. Sci. USA 74 (1977) 319-323.

33. Hart, R.W. Role of DNA repair in cell aging. In "Aging, Carcinogenesis and Radiation Biology". Editor, K.C. Smith, Plenum Press, New York (1976).

34. Reznikoff, C.A., J.S. Bertram, D.W. Brankow and C. Heidelberger. Quantitative and qualitative studies of chemical transformation of cloned C3H mouse embryo cells sensitive to postconfluence inhibition of cell division. Cancer Res. 33 (1973) 3239-3249.

35. Di Paolo, J.A., P.J. Donovan, and R.L. Nelson. Transformation of hamster cells in vitro by polycyclic hydrocarbons without cytotoxicity. Proc. Nat. Acad. Sci. USA 68 (1971) 2958-2961.

36. Huberman, E. and Sachs, L. Cell susceptibility to transformation and cytotoxicity by the carcinogenic hydrocarbon benzo (A) pyrene. Proc. Nat. Acad. Sci. USA 56 (1966) 1123-1129.

37. DeMars, R. and J. L. Jackson. Mutagenicity testing with human cells. J. Environ. Pathol. Toxicol., in press (1977).

FUSION OF CYTOPLASTS AND KARYOPLASTS AND HOW THIS PERTAINS TO RESEARCH OF AGING AND TRANSFORMATION

Audrey L. Muggleton-Harris

Department of Medical Microbiology, Stanford University

School of Medicine, Stanford, CA 94305

INTRODUCTION

Before describing a method for fusion of isolated karyoplasts and cytoplasts of the human diploid cell WI-38, I would like to review some of the previous techniques and research involving cell hybridization and nuclear transfer. After this introduction I will suggest ways in which I feel the technique which I have developed in the laboratory of Dr. L. Hayflick at Stanford University can overcome many of the drawbacks found with the earlier techniques, as well as ways in which this particular approach can be utilized to study aging and malignancy.

Cell Hybridization

The technique of cellular hybridization using inactivated Sendai virus has demonstrated that aging and malignancy can be studied at the cellular level. Harris and his colleagues found that malignancy could be suppressed when tumor cells were fused to "normal" fibroblasts as well as L cell derivatives; however, it was observed that chromosome loss resulted in a greater frequency of malignant segregants (1). The genetic analysis of these cells was hampered by the heterogeneity karyotype and the identification of the malignant chromosomes was not accomplished at this time. A later study by this same group found that specific chromosomes were responsible for suppressing the malignant state and when these chromosomes were lost the malignancy would be re-expressed (2). Similar studies have been reported using HeLa cells and normal diploid fibroblast hybrids; the suppression of malignancy obtained was measured by tumor formation. Malignant;malignant hybrids retained their malignant characteristics, whereas fusions of malignant and normal cells resulted in suppression of malignancy (3). Hybridization of cells has also been useful in the study of cellular aging. Heterokaryons of senescent and young diploid fibroblast-like cells demonstrated dominance of the senescent trait with respect to DNA synthesis (4). However, reactivation of dormant chicken erythrocyte cell nuclei when fused to "young" WI-38 fibroblast cells as opposed to dormant nuclei fused to "senescent" fibroblasts (5) indicates that the "young" cell can reactivate an "old" nucleus, and thus senescence could be recessive.

The ability to enucleate cells with the fungal metabolite Cytochalasin B has enabled this type of study to be carried a stage further. Whole cells could be fused to enucleated cells - thus, a nucleus was placed into a different intracellular mileau consisting of 2 (or more) cytoplasms. Wright and Hayflick used mass populations of human cells which were fused to anucleate cytoplasms and antimetabolites to select the experimental fusions (6,7). The results of this work indicated that the aging phenomenon of the WI-38 fibroblast was under nuclear control. The authors did suggest, however, that the results were not unequivocal and that further studies should be pursued to clarify the role of the nucleus and cytoplasm in cellular aging.

Nuclear Transfers

Aging has also been studied by the transfer of individual nuclei from a cell into a different cytoplasmic environment. Hammerling's early studies with Acetabularia (8) have been extended recently by Berger and Schweiger (9). Their findings demonstrated that nuclear behavior could be influenced by the cytoplasm. Similar results were obtained using Amoebae (10) and Amphibia (11). One example of this approach concerns some early work which I pursued with J. F. Danielli. Although the unicellular protozoan Amoebae proteus was used for this study, the results we obtained in that work appear to be of relevance to some recent studies using the human diploid fibroblast.

The normal logarithmically-growing Amoebae has an infinite life span when cultured in the laboratory. By restricting the amount of food available to this organism, an alternative state of replication and finite life span can be obtained. Limiting the amount of food prevented division and the cell remained in this state of non-division for varying periods of time. When abundant food was once more made available to the organism, division took place, but a finite life span of the daughter cells derived from the experimental cell was observed in all cases. Two types of

Table 1. Protozoan nuclear transfers

Number of transfers	Survivors at 3rd day	% 3-day survivors dividing	Division type
Normal nucleus -- spanned cytoplasm			
230	112	100	All spanned
Spanned nucleus -- normal cytoplasm			
370	225	100	All spanned

division were noted prior to the death of the cells: 1) stem-line division where one of the two daughter cells always died and finally, after several divisions, both cells died, and 2) a "normal" logarithmically-dividing clone which reached significant numbers before all of the cells died. Thus the experimental cells had a finite life span which resulted in the two types of division observed. Nuclear and cytoplasmic transfers were made between "non-spanned" and "spanned" cells and some of the results are seen in Table 1. The conclusions gathered from these experiments was that the cell is able to select an alternative state of existence if specific external and internal conditions are imposed upon it (in this case limitation of food and prevention of division). The finite life span and subsequent behavior of these spanned cells was controlled by both nucleus and cytoplasm, each controlling a different form of multiplication. In the case of the cytoplasmic control, only a very small amount of "finite life span" cytoplasm was needed to alter the behavior of a normal cell.

Turning attention now to a differentiating cell system, it has been shown that the fully differentiated aged amphibian lens epithelial cell nucleus, when transferred into a parthogenetically activated and enucleated frog egg, is capable of producing a metamorphosing tadpole and small frog (11) (Table 2). Thus we see that in this particular case the genes for differentiation and early development were not lost or permanently inactivated by age or differentiation. Similar results have been found using other types of differentiated cells (12,13).

MANIPULATION AND FUSION OF TISSUE CULTURED CELLS

Due both to the size and to the necessity of maintaining sterility, the micro-manipulation of the cultured cell is both time-consuming and tedious. The elegant experiments of Elaine

Table 2. Results of amphibian nuclear transfers 7 day organ cultured adult lens used for donor cells. Leibovitz medium.

		% of total transfers			Later development as % of complete blastula			
Source of nuclei	No. of transfers	Uncleaved	Partial St. 8	Complete St. 8	Stage 10-11	Stage 12-24	Stage 18-20 (lens area apparent)	Stage 21-24 (cornea transparent)
Original stimulated lens epith. cells	136	84	8	2	10	0	0	0
First serial transfer from blastula	114	27	34	39	35	6	0	5
Second serial transfer from blastula	118	34	38	26.5	15	52.5	3	0

Diacumakos (14) who micromanipulates individual chromosomes of HeLa and embryonic lung cells are the type of experiments which could if pursued contribute valuable information in the areas of malignancy and aging. However, I decided to compromise and use the inactivated Sendai virus to achieve fusion of individual nuclei and enucleated cells.

Fig. 1 is a diagram showing the method by which reconstituted cells derived from enucleated cells and isolated nuclei were prepared. The complete details for this technique are reported in a manuscript submitted for publication (15), but following is a brief description of the steps which were followed. To prepare the cytoplasts, a single cell suspension of WI-38 diploid fibroblast cells was seeded onto a 22mm plastic disc and incubated overnight (Fig. 1a). Discs were then inverted in a centrifuge tube containing 8 mls of 10 μg/ml cytochalasin B and centrifuged at 12,000 RPM for 15 min at 37°C (Fig. 1b). After recovery from the centrifuge tube, the discs were placed into the incubator for 15 min. Two to five enucleated cells were selected and ringed with a glass cloning ring secured to the disc with sterile silicone grease. Each cloning ring enclosed a known number of cytoplasts and their location within the cloning ring was noted (Fig. 1c). The nuclei (karyoplasts) were prepared by seeding the plastic discs with twice the concentration of cells (approx 2,500), cultured in an incubator overnight, and the discs were then suspended in the cytochalasin B solution above a 1 ml layer of bovine albumin. This was then centrifuged at 12,200 RPM for 12 min at 37°C. Contaminating cell fragments and cytoplasts were retained in the upper level of the albumin and the karyoplast pellet formed at the bottom of the tube (Fig. 1d). The karyoplast pellet was then removed, resuspended in 0.25% trypsin in BME prior to seeding onto the cytoplasts, and measurements of the karyoplasts were made with a Filer micrometer eyepiece before being seeded onto the cytoplasts. Fusion was accomplished with the aid of inactivated Sendai virus (2000 HAu/ml); this was pipetted onto the cytoplasts, and the petri dishes containing the cytoplast/virus preparation were refrigerated at 4°C for 15 min. The karyoplast suspension was then added to the cytoplasts (Fig. 1e) and the dishes returned to 4°C for a further 15 min to allow for adsorption of karyoplasts to cytoplasts; they were then placed in a 6 percent CO_2 environment at 37°C for 60 min to encourage fusion (16). After 60 min, medium was pipetted slowly onto the dish and a further incubation at 37°C for 45 min was allowed. The medium was then removed and conditioned medium was added to the dish, which was then placed in an incubator with a CO_2 environment for 24 hr. This fusion procedure is similar to the one described by Poste (17).

Examination of the experimental cells after 24 hr would reveal a picture similar to that shown in Fig. 1f: complete fusions where nuclei were attached to the cytoplasts, partial fusions which were

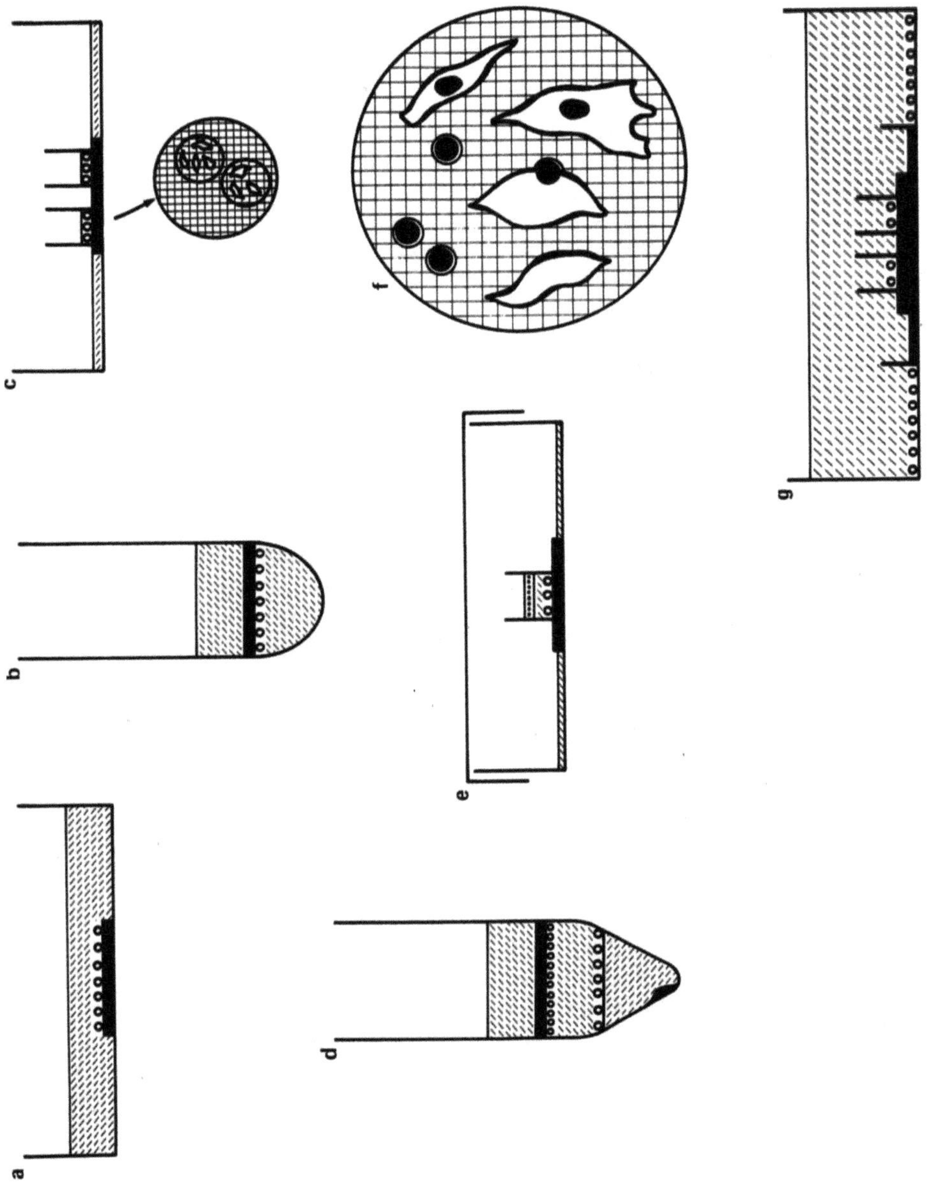
a
b
c
d
e
f
g

Fig. 1a. A 22 mm round plastic disc with grid markings seeded with approximately 1,200 cells (1) in a 60 mm petri dish containing BME + FCS (2). The dish was incubated at 37°C for 18 hr prior to experimental treatment.

1b. 50 ml round bottom centrifuge tube with inverted 22 mm plastic disc plus cells (1) immersed in 8 ml of 10 μg/ml cytochalasin B in BME (2) and spun at 12,000 RPM (17,300 G) for 12 min at 37°C to produce enucleated cells (cytoplasts).

1c. Gridded disc after centrifugation with cloning rings in place and enclosing selected cytoplasts (1). Diagram showing the view obtained within the interior of the cloning ring prior to the addition of inactivated Sendai virus (2).

1d. A 50 ml conical centrifuge tube with 1 ml of albumin at the bottom (1) and 13 ml of cytochalasin B (10 μg/ml) layered above (2). The inverted disc seeded with approximately 2,500 cells lodged within the cytochalasin B layer (3). The tube is centrifuged at 12,200 RPM (18,050 G) for 12 min at 37°C. The enucleated cells remain on the disc, cytoplasmic cell fragments are suspended at the interface of albumin and cytochalasin B solutions (4). The karyoplasts (nuclei) are pelleted at the bottom of the albumin layer (5).

1e. Disc with cytoplasts (1) after introduction of inactivated Sendai virus (2) and suspension of karyoplasts (3). The adsorption process takes place at 4°C and fusion at 37°C.

1f. Enlarged diagram of area within cloning ring 18-24 hr after the fusion procedure. The diagram shows (1) reconstructed cells or those cytoplasts and karyoplasts which have completely fused and appear similar to control cells (2) partial fusions where the karyoplast has only partially fused to the cytoplast membrane, and (3) cytoplasts and isolated karyoplasts which have not come into contact with one another and will disappear after 2-4 days.

1g. Disc plus experimental fusions still encircled by the cloning rings (1) and resting in a 35 mm petri dish (2) which itself sits on top of an irradiated feeder layer of cells (3). The non-dividing feeder cells condition the medium (4) in contact with the reconstructed cells but remain isolated from them.

cytoplasts that had a nucleus attached on the side, and isolated karyoplasts and cytoplasts. The exact location and condition was noted and a daily examination was made under phase contrast optics. Fusions were made between isolated nuclei and cytoplasts from similar population doubling levels (homotransfers) and also by combining nuclei and cytoplasms from different population doubling levels (heterotransfers); this technique enabled us to study aging at a cellular level by nuclear transfer.

Results of Nuclear Transfer Using Human Cells

The finite life span of the WI-38 in culture is a phenomenon which is well documented (18,19,20). The growth curve for the cells is shown in Fig. 2. As you can see, the senescent Phase III cells are committed to die, after an extended period of non-division. The Phase II cells on the other hand are dividing logarithmically and have many more population doublings ahead of them - that is, if something extra- and/or intracellular doesn't cause an alteration of their behavior. By isolating nuclei (karyoplasts) from Phase II cells (y) and fusing them into Phase III cytoplasms (o), one could accomplish the heterotransfer of a young nucleus into an aged cytoplasm (Oc/Yk). The isolation of young nuclei (Yk), and fusion into young cytoplasm (Yc) would be a homotransfer. Thus one could carry out the same type of nuclear transfers on the human cell which have been carried out in other systems. The results of the karyoplast or cytoplast fusions and replication of the nuclear transfers are shown in Tables 3 and 4. As Table 3 indicates, the number of fusions of nuclei to cytoplasm is relatively low in comparison with the original starting figures; these numbers are improving with recent experimentation. One factor which does emerge from Table 3 that could prove of interest in the study of aging and transformation is that the cytoplasts from the "old" culture always produced more "complete" fusions than the "young" cytoplast. Could this indicate a change with age of the cytoplast membrane, and is it one which encourages fusion of cellular membranes? The micromanipulation and reconstruction of individual cells may be a method of exploring changes in the cell membrane and how these changes may be influenced by intra- or extracellular environments.

Table 4 shows that the complete cells derived from the homo- and heterotransfers behaved differently from each other in number of replications and in their ability to produce viable clones of cells. Fig. 3 is a clone of cells derived from a Oc/Yk cell; it divided a limited number of times. Fig. 4 shows an area of cells derived from a Yc/Yk cell which was capable of clonal growth and culture. The reverse heterotransfer of Yc/Ok appears to suggest that the cytoplasm of the young cell is unable to influence the Phase III nucleus sufficiently to overcome its normal senescent behavior.

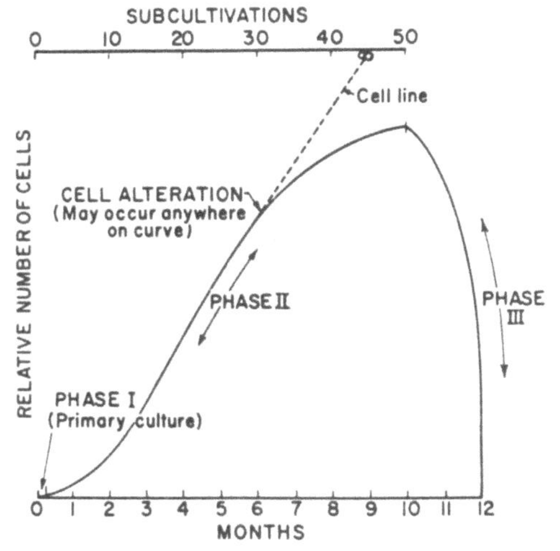

Fig. 2. Finite life span curve of human diploid fibroblasts WI-38. (Rep. from L. Hayflick and R. Moorhead, Exp. Cell Res. 25, 1961.

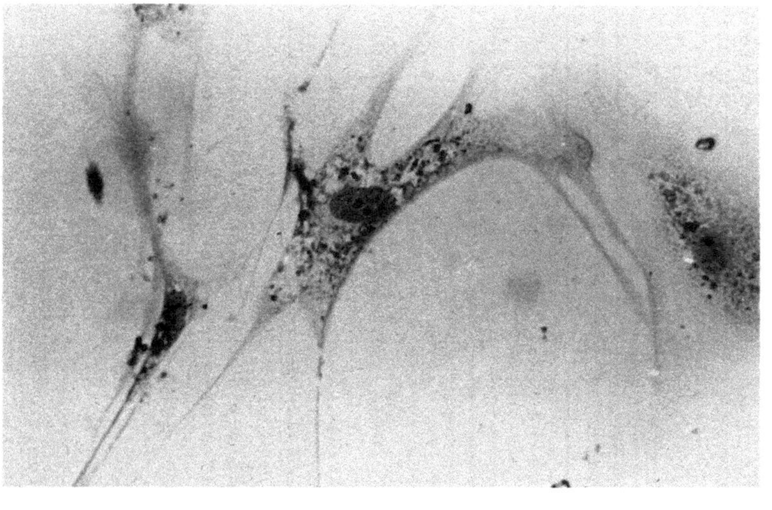

Fig. 3. A clone of cells (5) derived from a Oc/Yk reconstituted cell. This cell had divided and formed this clone within 14 days of fusion and had remained at this number for a further 40 days before cytoplasmic vacuolation and degeneration were noted.

Table 3. Fusion of Karyoplasts with Cytoplasts

Type of fusion and no. of cytoplasts selected	No. of cytoplasts with attached nuclei 5 hr after fusion	No. of "whole" cells 24 hr after fusion	No. of "complete" cells 4 days after fusion
Yc/Yk 1,962	291	112	71
Oc/Yk 1,563	383	259	208
Yc/Ok 626	161	42	22

Homotransfer Yc/Yk - Karyoplasts from PDL 12-24 into cytoplasts PDL 12-24

Heterotransfer Oc/Yk - Karyoplasts from PDL 12-24 into cytoplasts PDL 40-51

Heterotransfer Yc/Ok - Karyoplasts from PDL 40-51 into cytoplasts PDL 12-24

"Whole" cells are those fusions which have a nucleus within the cytoplasm and whose morphology is similar to that of a control cell. This evaluation is made 24 hr after fusion.

"Complete" cells are those cells which show no signs of nuclear degeneration or cytoplasmic vacuolation within 4 days of fusion.

Individual fusions of Yc/Yk always involved cytoplasts and karyoplasts from the same PDL, but the results combine data from several different PDLs.

Table 4. Replication of Reconstructed Cell

	No. of "complete" reconstructed cells 4 days after fusion	No. of cells dividing at least 2X	No. of cells dividing at least 3X*	No. of cells dividing at least 6X*
Yc/Yk	71	39	19	7
Oc/Yk	208	18	8	1
Yc/Ok	22	2	0	0
"Y" control cells	100	77	39	12
"O" control cells	100	48	21	4

Individual "O" with "Y" control cells were isolated and grown under similar conditions as experimental cells. However, no enucleation had taken place.

Many of the Yc/Yk combinations have replicated to form clones of 10^4 cells. The majority of the Oc/Yk cells ceased dividing after 3X but many of these cells remain alive 16 weeks after fusion. The Yc/Ok cells died 2 weeks after replicating.

* Those cells which had divided at least 6X were derived from cells which had divided 3X, and these in turn were derived from those cells dividing at least 2X.

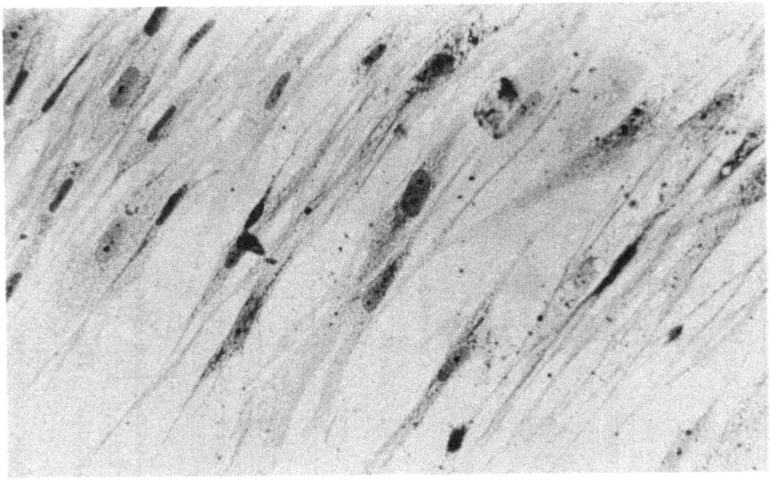

Fig. 4. A clone (approx 360) of cells derived from a Yc/Yk reconstituted cell 70 days after fusion.

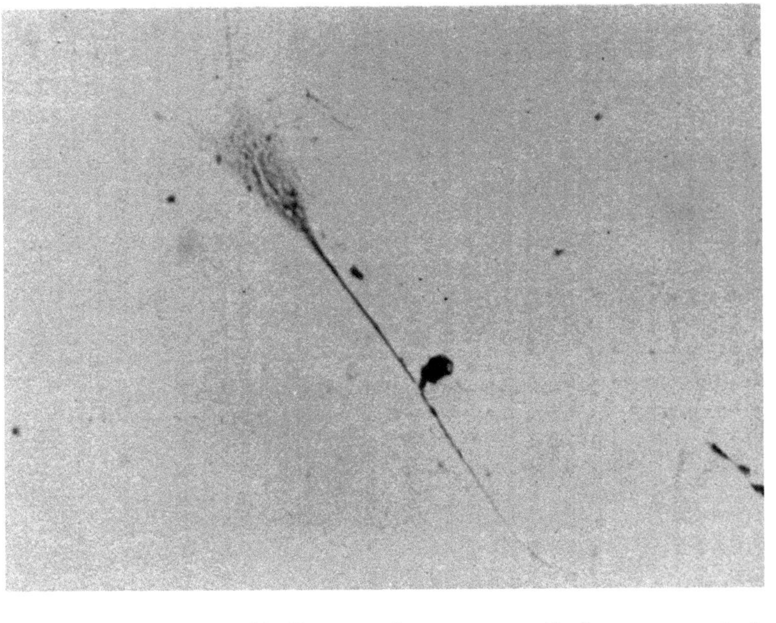

Fig. 5. An enucleated cell 5 min after removal from cytochalasin B. The nucleus is still attached by a short cytoplasmic stalk; if left, the nucleus would be pinched off after a few minutes.

PRESENT AND FUTURE EXPERIMENTS

The method of isolating nuclei through an albumin gradient does not encourage their viability. However, when individual nuclei were selected from the nuclear pellet (obtained when cytochalasin alone was used), or the karyoplasts were removed individually from the cytoplasmic filaments holding them after the enucleation procedure, greater numbers of complete cells and clonal growth were obtained.

Fig. 5 shows an enucleated cell which has the nucleus extruding on a cytoplasmic stalk. By using a micropipette attached to a deFonbrune microdissector, nuclei can be picked up and placed on the agglutinated surface of an enucleated cell. If cytokinesis were encouraged in the human cell, and specific chromosomes could be identified and confined into small "packages" in a manner similar to that published by Ege or Ringertz using animal cells (21), the "mini-cells" or small packages of chromosomes could be manipulated within the cell or enucleated with the aid of cytochalasin B and added to a whole cell which lacks these particular chromosomes. Thus the original work of Harris on malignancy could be carried further and preliminary experiments attempted to manipulate "aging" chromosomes.

Although I have not yet begun this particular approach, preliminary results in one area which is being pursued in our laboratory may give some information relevant to both aging and malignancy. In collaboration with Dr. T. Matsumura of our laboratory, individual nuclei have been isolated from a nuclear pellet obtained by enucleating HeLa cells and then fusing them into complete cells of very late PDL 70 Phase III WI-38 fibroblasts. These cells are at the end of their life span and have not divided for extended periods prior to the experiment. By the addition of a nucleus from the HeLa cell with its capacity for indefinite replication and malignancy, to a cell which has reached the end of its replication, a choice is given. Can the binucleate cell

1) utilize the chromosomes of the HeLa cell and assume the characteristics of indefinite replication and malignancy?

2) select for particular chromosomes from the donor nucleus and continue to replicate but not have malignant traits?

3) allow the fibroblast nucleus and/or cytoplasm to control the additional nucleus and the cell to retain its finite life span?

If the fibroblast cell is enucleated prior to fusion with the HeLa nucleus, further questions can be asked - is the cytoplasm of the nonmalignant cell capable of influencing the malignant nucleus and, if so, to what degree? Or is the nucleus of the HeLa cell capable of dominating the cytoplasm and working within it?

Preliminary results show that these binucleate cells appear to remain stationary as far as division for at least 28 days after fusion. The nucleus of the HeLa cell is very obvious and in some cases granular and cytoplasmic fragments are discarded from the cell, but after this period individual cells are seen to divide once more. Sometimes it appears that one cell has formed three or four cells. Nuclear material appears to fragment in some cells and is discarded. Those cells which did not receive a HeLa nucleus remained in a stationary phase. As these experimental heterokaryon cells are presently being cultured in the laboratory, it is not possible at this time to give actual figures on their replication, but the difference in behavior of the mononucleate and binucleate cells is quite apparent. Even though the senescent fibroblast cells divide several times after acquiring a HeLa nucleus, it cannot be said that these cells are transformed or malignant. Only if they replicated in sufficient numbers to enable tests for tumor formation, or replicated indefinitely and/or demonstrated a stable morphological change, could this statement be made. As these criteria have not been observed, I will confine my observations to our present results. The senescent fibroblast is capable of undergoing several replications after a HeLa nucleus is fused within the cell. The cells produced by this experimental fusion are fibroblast shaped and appear to have only one nucleus (Fig.6). The senescent WI-38 cells which did not receive an additional nucleus remained in a stationary state. Thus it would appear that the technique of cell assembly which I have described could be the means of studying cellular transformation and malignancy as well as aging.

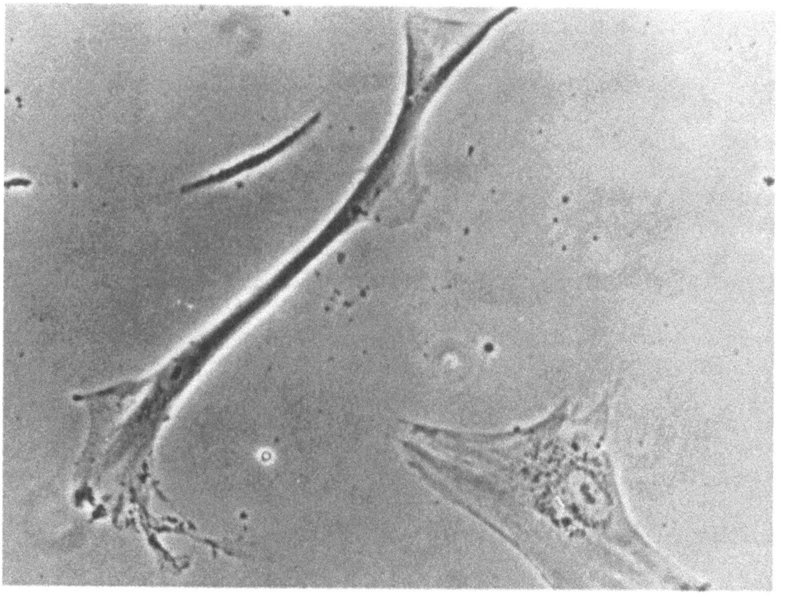

Fig. 6. Two cells from a clone of 5 cells derived from one senescent WI-38 cell (PDL 70) to which a HeLa nucleus had been fused. This photograph was taken 7 weeks after the fusion procedure.

ACKNOWLEDGMENTS

The cell assembly work on aging and transformed cells was supported by Research Grant HD-04004-06 from National Institutes of Health, Bethesda, MD, awarded to Dr. L. Hayflick. I wish to thank Dr. Hayflick for his enthusiastic encouragement of this experimental approach and to thank Nancy Pleibel for technical assistance.

REFERENCES

1. Klein WG, and Harris H. (1971) J Cell Sci 8:618

2. Klein WG, and Harris H. (1974) J Cell Sci 15:177

3. Stanbridge EJ. (1976) Nature (In press)

4. Norwood TH, Pendergrass WR, Sprague CA, and Martin EM. (1974) Proc Nat Acad Sci USA 71,6:2231

5. Rao NMV. (1975) 15th Ann Mtg, Amer Soc for Cell Biol

6. Wright WE, and Hayflick L. (1973) Proc Soc Exp Biol Med 144:487

7. Wright WE, and Hayflick L. (1975) Exp Cell Res 96:113

8. Hammerling J. (1953) Int Rev Cytol 2:475

9. Berger S, and Schweiger HG. (1975) J Cell Sci 17:517

10. Muggleton A, and Danielli JF. (1968) Exp Cell Res 49:116

11. Muggleton-Harris AL, and Pezzella K. (1972) Exp Gerontol 7:427

12. Gurdon JB, and Lasky RA. (1970) J Embryol exp Morph 24:227

13. McAvoy JW, and Dixon KE. (1974) J Exp Zool 189:243

14. Diacumakos,EG. (1973) Proc Nat Acad Sci USA 70,12:3382

15. Muggleton-Harris AL, and Hayflick L. (1976) Exp Cell Res (Submitted for publication)

16. Poste G. (1970) Microbios 2:227

17. Poste G, Reeve P. Methods in Cell Biology (ed DM Prescott)#7, p 228. Acad Press, New York (1973)

18. Hayflick L, and Moorhead PS. (1961) Exp Cell Res 25:86

19. Hayflick L. (1965) Exp Cell Res 37:614

20. Hayflick L. (1976) The cellular basis for biological aging. In Handbook on the Biology of Aging (eds C Finch, L Hayflick) Van Nostrand Reinhold, New York (In press)

21. Ege T, and Ringertz NR (1975) Exp Cell Res 94:469

DOMINANCE OR RECESSIVENESS OF THE SENESCENT PHENOTYPE IN SOMATIC CELL FUSIONS

Rosalind Yanishevsky and Gretchen H. Stein

Dept. of Molecular, Cellular and Developmental Biology

University of Colorado, Boulder, Colorado 80309

I would like to raise a few points for discussion of Dr. Muggleton-Harris' paper, after which I will describe some of our recent work that pertains to the central theme of this symposium. Dr. Muggleton-Harris' data suggest that both the cytoplasm and the nucleus of an old cell have the ability to confer senescence on a reconstructed cell. On the other hand, when Wright and Hayflick (1975) looked at (whole cell)/(cytoplast) fusions they concluded that the cytoplasm of an old cell does not have the ability to confer senescence on young cells. Can these results be reconciled? Let us consider several possibilities that could account for the apparent disparity between the two sets of data.

Individual cells in a WI-38 culture differ in their proliferative capacities (Cristofalo and Sharf, 1973; Absher *et al.*, 1974). In an old culture some cells will behave as if they are young with respect to proliferative capacity; likewise, in a young culture some cells will behave as if they are old. This heterogeneity of cellular proliferative capacity in both young and old cell cultures may have important consequences for the interpretation of these data because it is possible that cell fusions do not occur with equal frequency between young and old cellular components. For example, presumably cytoplasts, karyoplasts and whole cells from an old population are surrounded by similar old cell plasma membrane. Therefore, old cytoplasts might fuse more readily to old karyoplasts or to old whole cells, than to young karyoplasts or to young whole cells. In addition, Dr. Muggleton-Harris has suggested that with age there may be a change in the cytoplast membrane that renders it more susceptible to fusion. Thus, old cytoplasts might fuse more readily with old karyoplasts or old whole cells, than young cytoplasts would fuse with young karyoplasts or with whole young cells.

This preferential fusion hypothesis suggests that in Dr. Muggleton-Harris' experiment, old cytoplasts may have preferentially fused with old karyoplasts that exist in a young population. Thus, the cells reconstructed from a population of old cytoplasts and young karyoplasts may have behaved as old cells because both components were actually old, rather than because old cytoplasm confers senescence on a young karyoplast.

It may also be relevant that Muggleton-Harris performed fusions in the presence of trypsin to enhance the fusing capacity of the cells, whereas Wright and Hayflick (1975) did not. It is possible that trypsin enhances the fusion of old cells over young cells, for example, by revealing inherent surface differences between old and young cells. Therefore, the conditions of Muggleton-Harris' experiments may have favored the fusion of old karyoplasts in a young population, whereas this would not have occurred in Wright and Hayflick's experiments.

Obviously, to make any valid interpretations of the effect of heterogeneity of cellular proliferative capacity on the data, we need to know if there is a difference in fusion frequency among old and young cells and their cytoplasts and karyoplasts.

In the experiments of Wright and Hayflick (1975) cytoplasts were fused to whole cells treated with iodoacetate, which irreversibly inactivates a broad range of cellular components. The whole cells were treated with iodoacetate to inactivate their cytoplasms, while maintaining functional nuclei. In the cytoplasmic hybrids, or cybrids, the untreated cytoplasts rescued the treated whole cells and thus permitted survival of the cybrids.

Wright and Hayflick observed that old cytoplasts fused to iodoacetate-treated young whole cells yielded some proliferating hybrids; thus, they concluded that the cytoplasm does not control aging. The behavior of iodoacetate-treated cells is not well understood, and may lend some confusion to the interpretation of their results. For example, the amount of cytoplasm necessary to rescue an iodoacetate-treated cell might be less than that needed to express senescence. Occasionally, cells fuse and then pull apart after exchanging some of their cytoplasm (Harris _et al_., 1966). Without an appropriate cytoplasmic marker, it is impossible to determine whether this phenomenon occurred in these (cytoplast)/(whole cell) fusions.

Perhaps the ratio of old to young cytoplasm is important in determining senescence, and may account for the differences between the two sets of data. The cytoplast contains approximately 80% of the cytoplasm in a whole cell. The karyoplast contains approximately 10% of the cytoplasm. (Approximately 10% of the cytoplasm is lost during the enucleation procedure.) Therefore, the ratio of old to young cytoplasm is a 0.8:1 in an (old cytoplast)/(young whole cell) fusion whereas it is 8:1 in an (old cytoplast)/(young karyoplast) fusion. Perhaps the greater proportion of old cytoplasm to young cytoplasm in Dr. Muggleton-Harris' experiment accounts for its ability to confer senescence on a reconstructed cell containing a young nucleus. Unfortunately, we cannot properly weigh the importance of the ratio of old to young cytoplasm, because we do not know the precise effect of iodoacetate on the cytoplasm, particularly with regard to senescence determining factors.

In conclusion, it remains unclear as to whether the cytoplasm plays a role in determining senescence. Before we can reconcile the apparent disparity between the two sets of data, several questions need to be answered. For example, is there preferential cell fusion? What is the effect of iodoacetate treatment on the cells? Of course, other possibilities should be considered in reconciling these data. Perhaps we can explore these ideas further in the discussion session that follows.

Now I will turn to some of our experiments. We have heard earlier today that following fusion of a replicative young human diploid cell (HDC) with a nonreplicative senescent HDC, neither nucleus synthesizes DNA (Norwood et al., 1974). This may be interpreted to mean that nonreplicative senescent cells have a diffusible DNA synthesis inhibitor ("senescence factor") that is dominant over replicative normal nuclei. We have also heard that following fusion of a transformed human cell, namely HeLa or SV40 transformed human fibroblasts, with nonreplicative senescent HDC both nuclei synthesize DNA (Norwood et al., 1975). This may be interpreted to mean that transformed cells have a diffusible DNA synthesis inducer that is dominant over senescence factor.

Transformed cells such as HeLa differ from normal young HDC in at least several ways: 1) HeLa cells are immortal, 2) they are highly aneuploid, 3) they lack anchorage dependence (i.e., the requirement to attach and spread on a solid substratum in order to proliferate), and 4) they lack normal G1 growth regulation (i.e., the ability to cease DNA synthesis under conditions of low serum concentration in the medium or high cell density, by reversible arrest in the G1 phase of a cell cycle. In contrast, young HDC are 1) mortal, 2) diploid, 3) anchorage dependent, and 4) exhibit normal G1 growth regulation. We are interested in whether the ability to induce DNA synthesis in a senescent cell is correlated with immortality *per se*, or with another transformed property, such as the lack of normal G1 growth regulation.

To investigate this, we fused senescent HDC to T98G human cells, which are immortal yet exhibit normal G1 growth regulation. T98G cells were derived from a glioblastoma multiforma by Leonard Hayflick and Eva Pfendt. Gretchen Stein (1976, 1977) has characterized T98G cells and found that like normal cells, T98G cells exhibit stationary phase G1 arrest, have a high topoinhibition index (which is a measure of sensitivity to cell contact) and have a high serum requirement. However, like transformed cells, T98G cells are immortal, highly aneuploid with a modal chromosome number of 124-126, and anchorage independent. Thus, in T98G cells immortality is separated from a lack of normal G1 growth regulation.

The experimental procedure for the fusion of T98G and senescent HDC was as follows. For the HDC we used either WI-38 or IMR-90 cells. We used 1μm diameter polystyrene latex beads to label the cytoplasms of T98G cells and 2μm diameter beads to label the senescent cells. The beads were fed to the cells 48 hours before fusion. To discriminate between replicative and nonreplicative cells in the senescent cell population, we prelabeled the replicating cells with a high concentration of ^{3}H-thymidine (1 μCi/ml of S.A. 50 Ci/mM) for 48 hours before fusion. In this way, the replicative nuclei were identifiable as heavily labeled nuclei. In addition, such data on the proportion of cells that were labeled provide an estimate of the relative age of the population (Cristofalo and Sharf, 1973). Cells were fused using UV-inactivated Sendai virus. After fusion, a low level of ^{3}H-thymidine (0.01 μCi/ml) was added for 48 hours in order to label those nuclei that synthesized DNA after fusion. The cells were fixed and analyzed by autoradiography.

We measured several different parameters when analyzing the autoradiograms. There were three different labeling patterns in the cells: 1) heavily labeled nuclei, which are from replicative cells in the senescent population; 2) lightly labeled nuclei, which synthesized DNA after cell fusion; and 3) unlabeled nuclei, which did not synthesize DNA. The binucleate heterokaryons, exclusive of those containing a heavily labeled nucleus, were identified by the presence of at least 10 beads of both sizes and by the morphology of their nuclei. The morphology of the T98G and senescent HDC nuclei were sufficiently different to allow us to easily discriminate between them by staining them with Giemsa. The T98G nuclei almost always were round, stained darkly and contained more than 2 nucleoli. The senescent HDC nuclei almost always were oval, stained lightly and contained 1 or 2 nucleoli. The T98G and senescent HDC nuclei were scored by grain counts for whether one, or both, or neither of the nuclei had synthesized DNA. The fraction of labeled nuclei in the mononucleate cells was also scored.

The results in Table 1 show that 84% to 91% of the senescent nuclei in T98G-senescent HDC binucleate heterokaryons were unlabeled and, therefore, did not synthesize DNA. Thus, the ability to induce DNA synthesis in senescent nuclei is not correlated with immortality *per se*. The most obvious difference between T98G cells and cells such as HeLa, which can induce DNA synthesis in senescent nuclei, is that T98G cells exhibit normal G1 growth regulation whereas HeLa cells do not. Therefore, we hypothesize that the factor (which we call GR^- factor) that confers insensitivity to normal G1 growth regulation is the same factor that is necessary to induce DNA synthesis in senescent nuclei. Cells such as HeLa would contain GR^- factor, whereas cells such as T98G would lack GR^- factor. Of course, transformed properties other than lack of normal G1 growth regulation may be involved in the ability to induce DNA synthesis in senescent nuclei. We are testing this hypothesis by fusing senescent HDC with a series of cells with a variety of transformed properties.

Most, if not all, of the remaining 9 to 16% senescent HDC nuclei in the heterokaryons can be accounted for by assuming that they are replicative cells in the senescent population that did not synthesize DNA in the 48 hours prior to fusion (and hence were not heavily labeled), but did synthesize DNA in the 48 hours after fusion. This assumption is likely because a comparable fraction of the mononucleate senescent cells were also lightly labeled (see Table 1).

In Experiment I (Table 1) the majority (64%) of the binucleate heterokaryons showed that the T98G nucleus synthesized DNA, even though the senescent nucleus did not. This result suggested that the T98G nucleus was insensitive to the putative senescence factor. However, in two repetitions of this experiment, only 48% and 31% of the T98G nuclei synthesized DNA in heterokaryons containing nonreplicative senescent nuclei. At the same time, virtually all of the mononucleate T98G cells synthesized DNA. To explain these results, we hypothesize that in T98G nuclei, ongoing DNA synthesis is not inhibited by the putative senescence factor, although initiation of DNA synthesis is inhibited. Thus, we suggest that in Experiment I at least 64% of the T98G cells were synthesizing DNA at the time of fusion, whereas in Experiment III only about half as many cells were synthesizing DNA at the time of fusion. We are currently testing this hypothesis as well as investigating the mode of action of senescence factor in other mortal and immortal cell lines.

TABLE 1

Experiment	Percent of unlabeled senescent nuclei in T98G-senescent HDC binucleates[1]	Percent of nuclei in T98G-senescent HDC binucleates[1,2]				Percent of nuclei in mononucleates				
						senescent HDC			T98G	
		LL-UL[3]	UL-UL	LL-LL	UL-LL	LL[1]	UL[1]	HL[4]	LL	UL
I	84	64	20	16	0	14	86	58	91	9
II	91	48	43	7	2	2	98	34	100	0
III	90	31	59	10	0	9	91	24	99	1

[1]Cells with HL nuclei were excluded from these calculations.

[2]We recorded the labeling pattern for binucleates so that the T98G nucleus in on the left and the senescent HDC nucleus is on the right.

[3]Symbols are LL = lightly labeled, UL = unlabeled, HL = heavily labeled.

[4]Provides an estimate of the relative age of the cultures (Cristofalo and Sharf, 1973).

ACKNOWLEDGEMENTS

We gratefully acknowledge Doctors George Veomett and Lawrence Allred for invaluable discussions.

This work was supported by National Institutes of Health Grants AG00310 and CA13419, and American Cancer Society Grant NP156A.

REFERENCES

Absher, P.M., Absher, R.G. and Barnes, W.D.: Geneologies of Clones of Diploid Fibroblasts: Cinemicrographic Observations of Cell Division Patterns in Relation to Population Age. Exptl. Cell Res., 88:95-104, 1974.

Cristofalo, V.J. and Sharf, B.B.: Cellular Senescence and DNA Synthesis: Thymidine Incorporation as a Measure of Population Age in Human Diploid Cells. Exptl. Cell Res. 76:419-427, 1973.

Harris, H., Watkins, J.F., Ford, C.E. and Shoefl, G.I.: Artificial Heterokaryons of Animal Cells from Different Species. J. Cell Sci., 1:1-30, 1966.

Norwood, T.H., Pendergrass, W.R. and Martin, G.M.: Reinitiation of DNA Synthesis in Senescent Human Fibroblasts upon Fusion of Cells with Unlimited Growth Potential. J. Cell Biol., 64:551-556, 1975.

Norwood, T.H., Pendergrass, W.H., Sprague, C.A. and Martin, G.M.: Dominance of the Senescent Phenotype in Heterokaryons between Replicative and Post Replicative Fibroblast-Like Cells. Proc. Natl. Acad. Sci. USA, 71:2231-2235, 1974.

Stein, G.H.: Characterization of T98: A Polyploid Human Tumor Cell Line Showing Normal Growth Regulation in Vitro. J. Cell Biol., 70:24a, 1976.

Stein, G.H.: T98G: A Human Tumor Cell Line Which Has Acquired Normal Growth Regulation in Vitro. Manuscript in preparation, 1977.

Wright, W.E. and Hayflick, L.: Nuclear Control of Cellular Aging Demonstrated by Hybridization of Anucleate and Whole Cultured Normal Human Fibroblasts. Exptl. Cell Res. 96:113-121, 1975.

IMPROVED TECHNIQUES FOR THE FUSION OF MAMMALIAN CELLS

Richard L. Davidson

Division of Human Genetics, Children's Hospital Medical Center, and Department of Microbiology and Molecular Genetics, Harvard Medical School, Boston, Massachusetts 02115

As indicated by the title of this symposium "Is Senescence Dominant or Recessive in Somatic Cell Crosses", cell fusion has become an important technique for the study of the problem of cell senescence. However, despite the widespread use of cell fusion to study cell senescence and many other areas in mammalian cell biology and genetics, the induction of cell fusion has remained a major problem. In some cases, the ability (or inability) to induce fusion has been the limiting factor in experimental programs. The purpose of this paper is to present improvements in the techniques of cell fusion, which improvements could facilitate the types of experiments described in other papers in this symposium.

To date, the most widely used agent to induce cell fusion has been inactivated Sendai virus. The use of this agent, however, has been plagued by problems such as difficulties in preparing the virus and large variations in fusion inducing activity from one batch of virus to another. A number of attempts to find chemicals that are good fusing agents have been reported in recent years, but all of the agents tested have proven to have little fusion activity or to show a lack of reproducibility from laboratory to laboratory. Recent experiments on the induction of cell fusion by polyethylene glycol (PEG) suggest that this agent may solve many of the problems mentioned above.

The fusion inducing capacity of PEG was first demonstrated in experiments with plant cells (Kao and Michayluk, 1974). It was demonstrated soon thereafter that mammalian cells similarly could be induced to fuse by treatment with PEG (Pontecorvo, 1975).
In the experiments with mammalian cells, it was also shown that viable hybrids could be isolated following PEG induced fusion.

Experiments in our laboratory have confirmed that PEG induces the fusion of mammalian cells and can lead to the formation of viable hybrids. In the course of our experiments, we have analyzed a number of parameters involved in PEG induced fusion and have developed procedures for treating cells with PEG that are simple and effective (Davidson and Gerald, 1976; Davidson et al, 1976). The experiments and procedures described below deal only with the treatment of cells attached to a substrate. Techniques for the fusion of cells in suspension have not yet been as successfully worked out as the techniques for the fusion of cells in monolayer.

The optimal procedure for treating monolayer cells with PEG was developed in experiments involving the hybridization of mouse fibroblasts with rat glial cells. The mouse cells are of the line 3T3-4E and are resistant to the drug 5-bromodeoxyuridine; the rat cells are of the line RG6A-TgA and are resistant to the drug 8-azaguanine. Because of the enzyme deficiencies associated with their respective drug resistances, cells of neither line can grow in HAT medium. However, hybrids between the two cell lines can grow in HAT medium, which therefore can be used to select hybrids and determine the efficiency of hybridization induced by PEG.

Mixed cultures of mouse and rat cells were treated with PEG (at a variety of different molecular weights and concentrations) according to the protocol described below. PEG solutions were prepared by autoclaving the PEG and mixing it with warm medium (without serum). The concentrations of the PEG solutions are expressed on a weight/weight basis; for example, a 50% solution of PEG can be prepared by mixing 5 grams of PEG and 5 ml of medium.

The following protocol has been found to yield optimal results, as judged by the frequency of hybridization. However, it also appears to be equally well suited for the formation of multinucleate cells.

1. Inoculate 60 mm dishes with a mixture of the two parental cell types, with enough cells to give a semi-confluent monolayer. Allow the cells to incubate for several hours at 37°C to become firmly attached to the substrate.

2. Thoroughly aspirate the medium from the dishes.

3. Add 3 ml of PEG solution at room temperature to each dish. Let the dishes sit for one minute at room temperature.

4. Aspirate the PEG and rinse the dishes rapidly three times with medium. Do not attempt to remove all the residual PEG from the dishes.

5. Incubate the cells for several hours to allow the completion

of fusion. The cultures then can be trypsinized and the cells replated in HAT medium to determine the hybridization frequency.

Some of the variables tested had no apparent effect on the frequency of hybridization. For example, it did not seem to matter whether the cells were exposed to PEG at room temperature or at $37^{o}C$. The pH of the PEG solution also seemed to have relatively little importance. Other variables, however, were found to have a great effect on the frequency of hybridization. The length of exposure of the cells was found to be critical. With the cross between mouse fibroblasts and rat glial cells, one minute was the optimal time of exposure. With either shorter or longer exposures to PEG, the yield of hybrid cells dropped significantly. The decrease in the yield of hybrids with PEG exposures longer than one minute seems to be due to increased toxicity. In contrast, the decreased hybridization with PEG exposures shorter than one minute seems to be due to a drop in the extent of cell fusion. Other critical factors are the molecular weight and concentration of the PEG, as described below. With all of the conditions optimized, we have observed that as many as 5% of the parental glial cells can fuse with mouse fibroblasts to produce viable hybrids. This is comparable to the highest frequency of hybridization that has been obtained with Sendai virus.

The effects of changes in PEG concentration were investigated initially with PEG 6000. It became apparent very quickly that the concentration of PEG has a marked effect on the frequency of hybridization and that even a small change in PEG concentration could cause a large change in the yield of hybrids. For example, the optimal concentration of PEG 6000 (in terms of the hybridization frequency) was found to be 50%. When the concentration was decreased to 40%, the yield of hybrids decreased by a factor of 100 (to the level observed in untreated control cultures). Similarly, when the PEG concentration was increased to 60%, the yield of hybrids dropped by a factor of 10.

The decreased yield of hybrids with PEG concentrations lower than 50% seems to be due to a sharp drop in the extent of cell fusion. In contrast, the decreased hybridization with PEG concentrations greater than 50% seems to be due to an increase in the toxicity of the PEG. At the concentrations above 50%, the yield of hybrids decreases even though the degree of cell fusion actually seems to increase. Thus, the optimum PEG concentration for hybridization is apparently determined by a balance between two factors - cell fusion and cell viability.

The structure of PEG is $HOH_2C(CH_2OCH_2)_nCH_2OH$.

PEG is a linear molecule, and it can have different molecular weights according to the number of repeating subunits. Commercially, PEG grades with a wide range of molecular weights are available. We have tested the fusion activity of PEG grades with molecular weights ranging from 200(with approximately 3 repeating subunits/molecule)to 6000 (with approximately 150 repeating subunits/molecule). PEG 200, 300, 400, 600, 1000, 1540, 4000, and 6000 were tested, and all of them were found to be capable of inducing hybridization. However, there were significant differences among the grades, in terms of the optimum concentration for hybridization, the maximum hybridization efficiency, and the degree to which hybridization was affected by changes in concentration.

Of all the PEG grades tested, PEG 200 was found to be the least effective in inducing hybridization. The maximum hybridization efficiencies obtained with all the other PEG grades (300-6000) were at least 6 times greater than that obtained with PEG 200. The results suggest that, as the size of the PEG molecule decreases, there is a minimim molecular weight (around 300) or a minimum number of repeating subunits per molecule (around 5), below which the fusion inducing activity begins to decline. Above the minimum there seems to be relatively little difference in the maximum hybridization efficiencies obtainable with the different PEG grades (from 300-6000).

For all PEG grades from 300-6000, the maximum hybridization efficiency was achieved with concentrations of 50 or 55%. It seemed that the grades with molecular weights of 1000 or greater had their optimum concentrations at 50%, whereas the grades with molecular weights lower than 1000 had their optimum concentration at 55%. Among all the grades and all the concentrations tested, the highest absolute hybridization efficiency was observed with PEG 1000 at a concentration of 50%.

In addition to the variations among the grades in terms of the optimum concentration and the maximum hybridization efficiency, differences in the effects of changes in PEG concentration were also observed. For example, when PEG 6000 is diluted from 50% (its optimum concentration) to 45%, approximately 80% of the hybridization inducing activity is lost. In contrast, when PEG 1000 is diluted from 50% (also its optimum concentration) to 45%, only approximately 30% of its activity is lost.

At present, we cannot explain the effects on hybridization efficiency of changes in either PEG molecular weight or concentration. However, on the basis of the results presented above, it seems that PEG 1000 at a concentration of 50% represents the best choice for inducing cell hybrids. This choice is based on two reasons: (1) PEG 1000 at 50% produces a higher efficiency of hybridization than any other grade at any concentration; (2) PEG 1000 is less sensitive to changes in concentration (in the region of its optimum concentration) than are any other grades.

We have attempted to optimize the conditions of PEG induced hybridization of cells attached to a substrate. Although the conditions were optimized for one specific cross, they seem to be applicable to most other cases in which fusion of attached cells is involved, both for hybridization and heterokaryon formation. Since cell fusion is being used extensively to study cell senescence, improvements in the techniques of cell fusion should facilitate these studies.

LITERATURE CITED

1. Davidson, R.L. and Gerald, P.S. Somatic Cell Genetics 2, 165-176, 1976.

2. Davidson, R.L., O'Malley, K.A., and Wheeler, T.W. Somatic Cell Genetics 2, 271-280, 1976.

3. Kao, K.N. and Michayluk, M.R. Planta (Berlin) 115, 355-367, 1974.

4. Pontecorvo, G. Somatic Cell Genetics 1, 397-400, 1975.

MAPPING THE SV40 INTEGRATION SITES IN SV40-TRANSFORMED HUMAN CELLS

Carlo M. Croce

The Wistar Institute of Anatomy and Biology
36th Street at Spruce
Philadelphia, Pennsylvania 19104

Since somatic cell hybrids between rodent cells derived from continuous cell lines and human cells lose (segregate) human chromosomes (Weiss and Green, 1967), it becomes possible to map human genes to specific human chromosomes (Ruddle, 1973) by determining whether the expression of a given human phenotype segregates concordantly with the presence of a specific human chromosome in the hybrid cells. Cells derived from different mammalian species, including man (Girardi et al., 1965), can be transformed by the small DNA tumor virus, simian virus 40 (SV40). The SV40 genome becomes covalently linked to cellular DNA in these cells (Sambrook et al., 1968), which also express a nuclear tumor (T) antigen (Black et al., 1963), apparently coded by a viral gene (Tegtmeyer, 1974), and a tumor specific transplantation antigen on the cell surface (Habel, 1965). Fusion of T antigen-positive with T antigen-negative cells results in the production of heterokaryons which express T antigen in the nuclei derived from both parent cells (Steplewski et al., 1968) and in the production of T antigen-positive and negative hybrid cells (Weiss, 1970).

Since mouse and human chromosomes can be identified in the hybrid cells by either quinacrine (Caspersson et al., 1970) or Giemsa (Seabright, 1971) banding staining, we decided to study the expression of SV40 T antigen in hybrids between mouse cells and different SV40-transformed human cells to localize the chromosomal site of integration of SV40 in these cell lines and to determine whether SV40 is integrated in only one or more than one chromosome.

TABLE 1

Expression of SV40 T antigen in mouse-human hybrids

		All hybrids		
		+	C-7	-
SV40 T antigen	+	71		0
	-	3		12

TABLE 2

Expression of SV40 T antigen in mouse-human hybrid subclones

Hybrid clones	No. of subclones or triple hybrids	No. of T antigen-positive subclones or triple hybrids
52-58 Cl 19	15	3
52-62 (1) Cl 5 BrdU	9	7
52-62 (1) Cl 16 BrdU	5	4
52-62 (1) Cl 5 BrdU x IR	20	12

TABLE 3

Correlation between the expression of SV40 T antigen and the presence of human chromosome 7 in hybrid subclones and triple hybrids

		Human chromosome 7	
		+	-
SV40 T antigen	+	26	0
	-	0	23

Expression of SV40 T antigen in mouse-human hybrids

A series of hybrid cell clones between thymidine kinase-deficient Cl 1D mouse cells and cells derived from two different SV40-transformed cells (LN-SV and W18Va2) were tested for the expression of SV40 T antigen and studied for the presence of human chromosomes. As shown in Table 1, concordant segregation of the expression of SV40 T antigen and the presence of human chromosome 7 was observed (Croce *et al.*, 1973). Three of the 86 clones analyzed retained human chromosome 7 and did not express SV40 T antigen, suggesting that the genome of SV40 is integrated in only one of the parental chromosomes 7 derived from the LN-SV transformed fibroblasts. To confirm these findings three T antigen-positive clones were subcloned. One of these clones [56-62 (1) Cl 5 BrdU] contained only one human chromosome, the 7, and had been counterselected in medium containing 90 ug/ml of 5-bromodeoxyuridine (BrdU). 56-62 (1) Cl 5 BrdU cells were also hybridized to mouse IR cells deficient in hypoxanthine-phosphoribosyltransferase (HPRT) and the triple hybrids were selected in hypoxanthine-aminopterin-thymidine (HAT) medium (Littlefield, 1964). As shown in Table 2, the hybrid subclones and triple hybrids segregated into SV40 T antigen-positive and T antigen-negative cells. Concordant segregation of the expression of SV40 T antigen and the presence of human chromosome 7 was observed in all hybrids which were studied (Table 3). These experiments confirm that the gene for SV40 T antigen is located in human chromosome 7 and rule out the possibility that the expression of the SV40 T antigen in SV40 T antigen-positive hybrid clones was caused by a transfer of the viral genome to a mouse chromosome, since all the hybrid subclones which have lost the human chromosome 7, but retained all the mouse chromosomes, were SV40 T antigen-negative (Croce and Koprowski, 1974a).

Nuclei acid hybridization studies using these hybrid cells confirmed these results and indicated that one genome equivalent of SV40 DNA is integrated in human chromosome 7 (Khoury and Croce, 1975). We also wanted to determine whether the SV40 DNA integrates specifically in human chromosome 7. Therefore, we hybridized either mouse (IT-22) or Chinese hamster cells (A3) deficient in thymidine kinase with cells derived from a different SV40-transformed cell line (GM54VA). As shown in Tables 4 and 5, the IT-22 x GM54VA and A3 x GM54VA hybrids segregated into T antigen-positive and T antigen-negative hybrid clones. Counter-selection of the hybrid clones in medium containing BrdU resulted in the growth of hybrid cells which have concordantly lost the expression of SV40 T antigen and human chromosome 17 (Table 5) (Croce, 1977). Therefore, it can be concluded that SV40 DNA can be integrated in at least two distinctive chromosomal sites in SV40-transformed human cells: one located in human chromosome 7 and the other in human chromosome 17.

TABLE 4

Expression of SV40 T antigen in hybrids

IT-22 and GM54VA cells

	Selected in HAT	
Hybrids clones	% T antigen-positive cells	No. of hybrid cells containing human chromosome 17 / total analyzed
55-24 Cl 3	90	15/15
55-24 Cl 5	1	17/18
55-24 Cl 6	90	21/21
55-24 Cl 7	30	16/18
55-61 Cl 3	90	23/23
55-61 Cl 5	90	22/22
55-61 Cl 13	1	16/16
55-61 Cl 19	90	19/20
55-61 Cl 31	90	18/18
55-61 Cl 32	75	20/20
55-61 Cl 35	90	19/20
55-61 Cl 36	1	19/21

TABLE 5

Expression of SV40 T antigen in hybrids between A3 and GM54VA

	Selected in HAT		Counterselected in BrdU	
Hybrid clones	% T antigen-positive cells	No. of hybrid cells containing human chromosome 17 / total analyzed	% T antigen-positive cells	No. of hybrid cells containing human chromosome 17 / total analyzed
56-11 Cl 1	90	15/15	ND*	ND
56-11 Cl 2	80	10/11	1	0/10
56-11 Cl 5	90	12/12	1	0/12
56-11 Cl 8	75	9/10	ND	ND
56-11 Cl 9	50	13/13	1	0/10
56-11 Cl 11	1	15/15	1	0/10
56-11 Cl 14	80	13/14	ND	ND
56-11 Cl 15	90	15/15	1	0/10
56-11 Cl 16	1	13/14	ND	ND
56-11 Cl 17	90	12/12	1	0/15
56-11 Cl 18	90	12/13	1	0/10
56-11 Cl 24	90	12/12	1	0/11
56-11 Cl 29	90	11/11	1	0/10
56-11 Cl 30	90	11/13	ND	ND
56-11 Cl 31	1	12/12	ND	ND

*ND: Counterselection of the HAT-selected clones in medium containing BrdU was not done.

Expression of the transformed phenotype in somatic cell hybrids between normal mouse cells and SV40-transformed human cells.

To demonstrate that either human chromosome 7 or human chromosome 17, carrying the SV40 genome code for "transforming factors" responsible for the expression of the transformed phenotype, we decided to hybridize mouse peritoneal macrophages (MPM), which are nondividing cells (Epifanova and Terskikh, 1969) with either HPRT-dificient LN-SV or GM54VA SV40-transformed human cells. If human chromosomes 7 or 17 with the integrated SV40 genome are responsible for the unlimited growth of the human transformed cells and for the expression of the transformed phenotype, we should expect to obtain hybrids between mouse macrophages and SV40-transformed human cells that grow indefinitely in culture and that are transformed when the human chromosome carrying the SV40 genome is present.

No hybrids were obtained by fusing either LN or GM54 normal diplois human cells from which LN-SV and GM54VA were derived respectively and MPM (Croce and Koprowski, 1974b). On the contrary, hybrids were obtained by fusing MPM with either LN-SV or GM54VA human cells. As shown in Table 6, all hybrids between MPM and LN-SV cells were SV40 T antigen-positive and retained human chromosome 7. Twenty-seven MPM x GM54VA hybrids were also studied and found to be SV40 T antigen-positive and to contain human chromosome 17 (Croce, 1977). All these hybrid clones were found not to express the density dependent inhibition of growth (one of the characteristics of normal cells) and to grow in semisolid media (one of the characteristics of malignant cells) (Croce and Koprowski, 1974b).

These experiments indicate that the growth of the hybrid cells depends on the presence of the human chromosome carrying the SV40 genome and that the presence of this chromosome is responsible for the transformed behavior of the hybrid cells.

Since the human chromosomes carrying the SV40 genome are selectively retained in hybrids between MPM and SV40-transformed human cells, this type of hybrid appears to be extremely useful for mapping the integrated SV40 genome in different SV40-transformed human cells.

Expression of malignancy in somatic cell hybrids between MPM and SV40-transformed human cells.

To determine whether hybrid cells containing the human chromosomes carrying the SV40 genome are tumorigenic, we injected athymic nude mice, in which heterotransplantation of human tumors can be

TABLE 6

Expression of the transformed phenotype in hybrids between mouse peritoneal macrophages and LN-SV

Cells	SV40 T antigen *	Human chromosomes present in the hybrid clones**
Mouse peritoneal macrophages	-	
LN-SV	+	
Hybrids:		
Cl 1	+	4,6,7,11,12,14,17,19,20,X
Cl 2	+	4,6,7,11,12,17
Cl 6	+	6,7,17,20
Cl 7	+	4,5,7,11,12,17,18,20
Cl 11	+	4,6,7,11,12,17
Cl 12	+	4,7,11
Cl 13	+	7
Cl 17	+	6,7,11,12,14,17
Cl 19	+	1,4,6,7,11,12,17,19,20,21
Cl 23	+	7,12
Cl 28	+	4,5,7,11,12
Cl 35	+	2,4,7,11,12
Cl 41	+	7
Cl 46	+	4,7,11,12,20,21
Cl 47	+	4,6,7,11,12,14,19,20,X
Cl 61	+	4,5,7,11,12
Cl 62	+	7
Cl 63	+	4,6,7,10,11,12,14,17,20,22,X
Cl 75	+	4,7,11,12,17,19
Cl 81	+	7
Cl 82	+	7,11,12,17
Cl 87	+	7,12

* 100% of the cells of each hybrid clone were T antigen-positive. A minimum of 2,500 cells per each independent clone was studied for the presence of SV40 T antigen.

** A minimum of 20 metaphases per each clone was photographed and karyotyped. All the metaphases contained the human chromosome 7. The other human chromosomes were present in at least 5% of the metaphases analyzed per each clone.

successfully achieved (Rygaard and Paulsen, 1969), with MPM x LN-SV and MPM x GM54VA hybrid cells. The MPM x LN-SV hybrid cells used in this study contained 14 different human chromosomes (Table 7). The injection of the hybrid cells resulted in the formation of tumors. Five of those tumors were removed and transferred to tissue cultures. Table 8 summarizes the results of the karyologic analysis of the tumor lines. Human chromosome 7 carrying the SV40 genome was retained by all the tumorigenic cell hybrids (Table 8). Since the full complement of mouse chromosomes derived from the normal mouse parent was retained by the tumorigenic hybrid cells (Croce et al., 1975a), these results contrast with the hypothesis formulated by Harris et al. (1969) and Klein et al. (1971), who postulated that hybrids between normal and tumorigenic cells behave as normal cells.

To investigate whether hybrid cells which contain a diploid complement of mouse chromosomes are tumorigenic, we also injected nude mice with two different BALB/c MPM x LN-SV hybrid clones (10^7 cells/mouse). One (series A) contained a near diploid complement of mouse chromosomes and only human chromosome 7 (53-87-3 Cl 10), and the other (series B) contained a near diploid complement of mouse chromosomes and human chromosomes 7, 5, and 11 (53-87-3 Cl 43). Twenty mice out of twenty-two injected developed tumors. Two tumors of series A and five tumors of series B were transferred to tissue culture and the resulting cell lines were studied for the expression of SV40 T antigen and the presence of human chromosomes. As shown in Table 9, all the tumor lines were SV40 T antigen-positive and contained human chromosome 7 (Croce et al., 1975b). In the case of tumor lines Nulo IT-1, Nu10 IT-2 and Nu43, Nu43 IT-1, a single human chromosome 7 per cell was observed in the great majority of the cells analyzed. No loss of any mouse chromosome was observed. Therefore, it can be concluded that the presence of a single copy per cell of human chromosome 7 carrying the SV40 genome is sufficient for the expression of tumorigenicity of the hybrids.

Somatic cell hybrids between either BALB/c or C57BL MPM and GM54VA human cells which contain the entire complement of mouse chromosomes and only human chromosome 17 were also injected into nude mice. All twenty nude mice injected developed tumors. Karyologic study of the tumor cells indicated that they retained the human chromosome 17 carrying the SV40 genome and were SV40 T antigen-positive.

TABLE 7

Presence of different human chromosomes in hybrid cells between C57BL/6 peritoneal macrophages and LN-SV human cells

Human chromosome	Frequency in hybrids, %*	Human chromosome	Frequency in hybrids, %*
4	5.9	13	5.9
5	60.8	14	7.8
6	70.6	15	17.6
7	100**	16	2.0
9	5.9	17	60.8
11	54.9	18	3.9
12	23.5	20	5.9

* A total of 51 metaphases was analyzed.

** On the average the hybrid cells contained 2.3 chromosomes 7 per cell.

TABLE 8

Karyological analysis of cells recovered from tumors induced in "nude" mice by inoculation of hybrid cells

Source of tumors	Removal, days after inoculation	Percent of cells showing SV40 T antigen	Human chromosomes*				
			5	6	7	11	17
"Nude" 1	22	100	8/45	35/45	45/45	0/45	0/45
"Nude" 2	28	100	0/30	21/30	30/30	0/30	0/30
"Nude" 4	42	100	0/39	25/39	39/39	1/39	15/39
"Nude" 7	70	100	0/26	10/26	26/26	0/26	1/26
"Nude" 8	76	100	0/33	3/33	33/33	0/33	0/33

* Numbers of metaphases showing presence of this chromosome over total analyzed.

TABLE 9

Correlation of the expression of SV40 T antigen and the presence of human chromosome 7 in tumorigenic hybrid clones and tumor cells

Hybrid cells	T antigen-positive cells %	Human chromosomes (average No. per cell)			Average No. of rearranged chromosome 7 per cell	No. of mouse chromosomes per cell		No. of metaphases analyzed*
		5	7	11		Average	Range	
Series A								
53-87-3 Cl 10	100	0	1.0	0	0	41.9	39-44	25
Nu10 IT-1	100	0	1.3	0	0.6	41.8	38-44	49
Nu10 IT-2	100	0	1.0	0	0.1	42.0	40-43	27
Series B								
53-87-3 Cl 43	100	0.9	1.2	0.5	0	40.3	39-42	30
Nu43 IT-1	100	0.6	1.1	0.3	0.1	41.1	39-43	39
Nu43 IT-2	100	0	2.0	0	0	41.4	38-44	24
Nu43 IT-3	100	0.1	1.9	0	0.1	41.6	40-43	37
Nu43 IT-4	100	0.2	2.6	0	0.0	40.3	39-42	39
Nu43 IT-5	100	0.4	1.9	0.2	0.1	41.8	39-47	55

*Less than 10% of the cells of 53-87-3 Cl 10, Nu10 IT-1, and Nu10 IT-2 were near tetraplois, and these were not included in the computation of the averages. Less than 20% of the cells of 53-87-3 clone 43 and its tumor derivatives were near tetraploid, and these were not included in the computation of the averages.

CONCLUSIONS

The investigation of somatic hybrids between drug-resistant rodent cells and different SV40-transformed human cells has allowed the mapping of SV40 integration sites in different SV40-transformed human cells. These studies indicate that there are at least two sites for the integration of SV40: one located in human chromosome 7 and the other located in human chromosome 17. Additional SV40-transformed human cell lines should be hybridized with mouse cells to determine whether the SV40 genome integrates perferentially in certain human chromosomes.

The hybridization between MPM and SV40-transformed human cells appears to be the system of choice to map the SV40 integration sites, since human chromosome carrying the SV40 genome is selectively retained by the hybrid cells. However, it is necessary to hybridize the SV40-transformed human cells also to rodent cells derived from continuous cell lines to demonstrate that the loss of the chromosomes selectively retained by the MPM x SV40-transformed human cell hybrids results in the loss of the expression of SV40 T antigen.

These studies also indicate that in SV40-transformed cells the chromosomes carrying the genome of SV40 exercise a positive control of the transformed phenotype and are also responsible for the expression of tumorigenicity.

Study of somatic cell hybrids between normal mouse cells and human cells derived from different malignancies could also result in the identification of human chromosomes responsible for the expression of malignancy. Such studies are in progress using somatic cell hybrids between contact-inhibited mouse cells and human cells derived from fibrosarcomas and melanomas.

ACKNOWLEDGMENTS

This work was supported in part by USPHS research grants CA 10815 and CA 16685 from the National Cancer Institute, GM 20700 from the Institute of General Medical Sciences, and RR 05540 from the Division of Research Resources, and the National Foundation-March of Dimes (Basil O'Connor Starter Grant), and by VC-220 from the American Cancer Society. C.M.C. is a recipient of a Research Career Development Award (1K04CA00143) from the National Cancer Institute.

REFERENCES

Black, P. H., W. P. Rowe, H. C. Turner and R. J. Huebner (1963) A specific complement fixing antigen present in SV40 tumor and transformed cells. Proc. Nat. Acad. Sci. USA 50, 1148-1156.

Caspersson, T., L. Zech, C. Johansson and E. J. Modest (1970) Identification of human chromosomes by DNA-binding fluorescent agent. Chromosome 30, 215-227.

Croce, C. M. (1977) Assignment of the integration site for simian virus 40 to chromosome 17 in GM54VA, a human cell line transformed by simian virus 40. Proc. Nat. Acad. Sci. USA, in press.

Croce, C. M. and H. Koprowski (1974a) Concordant segregation of the expression of SV40 T antigen and human chromosome 7 in mouse-human hybrid subclones. J. Exp. Med. 139, 1350-1353.

Croce, C. M. and H. Koprowski (1974b) Somatic cell hybrids between mouse peritoneal macrophages and SV40-transformed human cells. J. Exp. Med. 140, 1221-1224.

Croce, C. M., A. J. Girardi and H. Koprowski (1973) Assignment of the T-antigen gene of simian virus 40 to human chromosome C-7. Proc. Nat. Acad. Sci. USA 70, 3617-3620.

Croce, C. M., D. Aden and H. Koprowski (1975a) Somatic cell hybrids between mouse peritoneal macrophages and simian virus-40-transformed human cells: II. Presence of human chromosome 7 carrying the simian virus 40 genome in cells of tumors induced by hybrid cells. Proc. Nat. Acad. Sci. USA 72, 1397-1400.

Croce, C. M., D. Aden and H. Koprowski (1975b) Tumorigenicity of mouse-human diploid hybrids in nude mice. Science 190, 1200-1202.

Epifanova, O. I. and V. V. Terskikh (1969) On the resting periods in the cell life cycle. Cell Tissue Kinet. 2, 75-93.

Girardi, A. J., F. Jensen and H. Koprowski (1965) SV40-induced transformation of human diploid cells: crisis and recovery. J. Cell Comp. Physoil. 65: 69-84.

Habel, K. (1965) Specific complement fixing antigens in polyoma tumors and infected cells. Virology 25, 55-61.

Harris, H., O. J. Miller, G. Klein, P. Worst and T. Tachibana (1969) Suppression of malignancy by cell fusion. Nature 223, 363-368.

Khoury, G. and C. M. Croce (1975) Quantitation of the viral DNA present in somatic cell hybrids between mouse cells and SV40-transformed human cells. Cell 6, 535-542.

Klein, G., U. Bregula, F. Wiener and H. Harris (1971) The analysis of malignancy by cell fusion. I. Hybrids between tumor cells and L cell derivatives. J. Cell Sci. 8, 659-672.

Littlefield, J. W. (1964) Selection of hybrids from mating of fibroblasts in vitro and their presumed recombinant. Science 145, 709-710.

Ruddle, F. H. (1973) Linkage analysis in man by somatic cell genetics. Nature 242, 163-169.

Rygaard, J. and C. O. Paulsen (1969) Heterotransplantation of a human malignant tumor in nude mice. Acta Pathol. Microbiol. Scand. 27, 758-760.

Sambrook, J., H. Westphal, P. R. Srinivasan and R. Dulbecco (1968) The state of viral DNA in cells transformed by SV40. Proc. Nat. Acad. Sci. USA 60, 1288-1295.

Seabright, M. (1971) A Rapid banding technique for human chromosomes. Lancet 2, 971-972.

Steplewski, Z., B. B. Knowles and H. Koprowski (1968) The mechanism of internuclear transmission of SV40-indiced complement fixation antigen in heterokaryocytes. Proc. Nat. Acad. Sci. USA 59, 769-776.

Tegtmeyer, P. (1974) Altered patterns of protein synthesis in infection by SV40 mutants. Cold Spring Harbor. Symp. Quant. Biol. 39, 9-15.

Weiss, M. C. (1970) Further studies on loss of T antigen from somatic cell hybrids between mouse cells and SC40-transformed human cells. Proc. Nat. Acad. Sci. USA 66, 79-86.

Weiss, M. and H. Green (1967) Human-mouse hybrid cell lines containing partial complements of human chromosomes and functioning human genes. Proc. Nat. Acad. Sci. USA 38, 1104-1111.

Discussion of Dr. Croce's Paper

Orlando J. Miller, M.D.

College of Physicians and Surgeons, Columbia University

701 West 168th Street, New York, New York 10032

Carlo Croce has presented very convincing evidence that SV40 viral expression and malignancy are dominant in mouse-human hybrid cells. He has shown that the virus integrated into chromosome 7 in three independently transformed human lines and into chromosome 17 in a fourth. Clearly there is not a single site of integration, and one wonders whether chromosome 7 (or 17, for that matter) has more than one site of SV40 integration. This question could be answered by mapping the integration site, much as one would map any other gene locus, by starting with cell lines in which chromosome 7 has been split into two parts by a translocation, or by inducing breaks in chromosome 7 by irradiating hybrid cells in which there is still a diploid mouse complement and the only human chromosome present is a number 7 carrying an integrated SV40 viral genome.

The site of integration of another oncogenic virus, a murine leukemia virus (MuLV), has been mapped by classical genetic methods in the genome of the AKR strain of highly inbred mice. At some point in the history of this strain, perhaps during its development from A strain mice by Jacob Furth, MuLV integration occurred. That is, in this case, viral integration occurred not in a cultured cell line but in a living mammal, and the virus has persisted in its integrated state, presumably without change of position, through many generations. Recently, Chatthopadhyay et al. (1975) have shown by molecular hybridization, that the MuLV virus is integrated at a specific site on mouse chromosome 7, earlier shown by this group to be about 12 map units from Gpi-1 and 30 map units from the c locus, with gene order centromere-Akv-1-Gpi-1-c.

MuLV can be integrated at other sites, e.g. Ak-2, and it is not at all clear whether the potential sites of integration are few or many for any particular virus. The integration of many bacterial viruses proceeds by a highly site specific mechanism in which specific nucleotide sequences play an important role. Integration of SV40 virus might take place by a similar mechanism, or by a less highly site-specific mechanism based on homology between a small number of sites in viral and host genomes, or even by a relatively non-site-specific mechanism. To investigate this question at a molecular level one could digest total hybrid cell DNA with HpaII or BamHI restriction endonuclease, which cleaves SV40 at a single site, separate the resultant DNA fragments by agarose gel electrophoresis and determine by in situ hybridization which fragments contain nucleotide sequences complementary to viral cRNA. This approach has been used to show that the nucleotide sequences in the SV40 virus and/or its site of integration in the host genome are different in each of five independently transformed BALB/c 3T3 mouse cell lines (Ketner and Kelly, 1976).

I want to say a few words about the question of whether viral expression is dominant or recessive in various systems. In Carlo's system, expression is dominant, as expected if the integrated virus makes viral RNA transcripts in an unregulated, or constitutive, fashion. However, cases are known in which viral expression is regulated by the product of another gene. For example, the Fv-1 gene of the mouse determines sensitivity to certain RNA tumor viruses. The Fv-1^b allele confers resistance to N-tropic MuLV while the Fv-1^n allele confers resistance to B-tropic virus. Inbred strains of mice are homozygous for one or the other of these alleles, but F_1 hybrids between mice that are Fv-$1^{n/n}$ and Fv-$1^{b/b}$ produce heterozygous Fv-$1^{b/n}$ individuals who are resistant to both B- and N-tropic viruses (Pincus et al., 1971). The Fv-1 gene does not act at the level of virus penetration of the cell, since vesicular stomatitus virus enters cells when given the coats of either N- or B-tropic viruses (Huang et al., 1973). The provirus integration step may also be intact (Sveda et al., 1974), although Tennant et al. (1974) have claimed that virus restriction occurs early after infection, possibly before integration. Jolicoeur and Baltimore (1976) showed that the amount of virus-specific RNA in the cytoplasm of resistant mouse cells (N-type virus in Fv-$1^{b/b}$ cells, say) was reduced 70-100 fold, and suggested that restriction by the Fv-1 gene occurs either at or just before the level of transcription or that the RNA transcripts are rapidly degraded after synthesis. Whatever the mechanism, the Fv-1 gene regulates MuLV production.

Fenyö et al. (1974) showed that somatic cell hybridization of A9 cells, which are Fv-$1^{n/n}$ and produce an N-tropic MuLV, with cells of either C57Bl or ACA strain mice, which are Fv-$1^{b/b}$, led to a marked reduction in MuLV production. Thus the dominance relations seen in germinal F_1 hybrids are found also in somatic cell hybrids. Interestingly, this marked reduction in C-type virus production was correlated with inability of the hybrid cells to grow progressively when transplanted into histocompatible hosts. While this association may be fortuitous, the proposed genetic model may still provide a simple way of thinking about dominance and recessiveness in such hybrids. Thus, in Carlo Croce's system, SV40 viral expression and malignancy are dominant because the oncogenic virus is expressed constitutively. On the other hand, in Henry Harris's system, the malignancy of Ehrlich ascites tumor, MSWBS methylcholantherene-induced sarcoma, SEWA lymphoma and A9HT fibroblasts was suppressed by fusion with normal diploid fibroblasts or other non-malignant or weakly malignant cells (Wiener et al., 1974a,b), with malignancy only reappearing if chromosomes derived from the normal parent were lost. Harris worked out a nice system to test the idea that loss of a specific chromosome was required. If this were so, then fusion of A9HT (malignant) cells with normal diploid cells should produce hybrids with, at best, a low level of malignancy unless both copies of the specific normal chromosome are lost. This idea can be checked by scoring a series of malignancy subclones that arise from such non-malignant hybrids. In doing so on a series of Henry Harris's hybrids we have found evidence suggesting that mouse chromosomes 4 and 7, possibly 6, carry such suppressors (D.A. Miller et al., unpublished). Further work (unpublished) in George Klein and Harris's laboratories, using near-diploid tumor lines instead of the heteroploid A9HT tumor line, has strengthened the evidence favoring a role for one of these chromosomes. It is possible, I suppose, that the transplantability of these hybrid cells is influenced by the C-type virus they produce, but whether or not that is the case, the same genetic model serves to clarify the situation. In short, heterogeneity among different cell lines or species is to be expected, with malignancy acting as a dominant when simple ("bacterial type") regulatory systems are irrelevant, and as a recessive when they are important.

Are these studies of malignancy, whether induced by virus, chemical carcinogen or other modality, relevant to the question raised in the title of this workshop: "Is Senescence Dominant or Recessive in Somatic Cell Crosses?" I think so, even if it is only because they provide model systems. One could go further, and suggest that in some cases aging is due to loss of an integrated virus, analogous to the lack of proliferative capacity of mouse peritoneal macrophages, which are genetically programmed not to divide, and their gain in proliferative capacity when they have gained the SV40 viral genome integrated into a human chromosome. In this case, the virus returned the cells to a 'neoplastoid' or stem cell state. In other cases, an environmental stimulus of a a different sort may be all that is required to do this. What I have in mind is the recent work of Mintz and Ilmensee (1975) in Philadelphia and Papaioannou et al. (1975) in Richard Gardner's laboratory at Cambridge. They found that mouse teratocarcinoma cells, whether maintained for up to eight years in an ascites tumor form or for over two years in cell culture, were able to turn their back on their neoplastic past and become perfectly normal when placed on the right environment: a developing blastocyst. The cells, homing in on the inner cell mass, participated in the differentiation of all kinds of tissues, including germinal tissues, and produced normal tetraparental mice. The cells even underwent meiotic division, producing normal haploid gametes and normal progeny. Mintz and Ilmensee (1975) concluded from their study that conversion to neoplasia need not involve structural changes in the genome (the teratocarcinoma cells remained diploid and euploid) but rather could involve simply a change in gene expression.

The proliferation capacity of cells can thus be decisively influenced by genes, viruses and other environmental factors. Their complex interaction in the processes of aging can be confidently predicted--and may be clarified by the kinds of studies now being used to understand other complex phenotypes.

References:

Chattopadhyay, S.K., Rowe, W.P., Teich, N.M., and Lowry, D.R.: Definitive evidence that murine C-type virus inducing locus Akv-1 is viral genetic material. Proc. Nat. Acad. Sci. U.S.A. 72: 906-910 (1975).

Fenyö, E.M., Nazerian, K., and Klein, E.: Characteristics of murine C-type viruses. II. The behavior of viruses resident in various cell lines and their hybrids on BALB/3T3 and mouse embyro fibroblast cultures. Virology 59: 574-579 (1974).

Huang, A.S., Besmer, P., Chu, L., and Baltimore, D.: Growth of pseudotypes of vesicular stomatitis virus with N-tropic murine leukemia virus coats in cells resistant to N-tropic viruses. J. Virol. 12: 659-662 (1973).

Jolicoeur, P., and Baltimore D.: Effect of Fv-1 gene product on synthesis of N-tropic and B-tropic murine leukemia viral RNA. Cell 7: 33-39 (1976).

Ketner, G., and Kelly, T.J., Jr.: Integrated simian virus 40 sequences in transformed cell DNA: analysis using restriction endonucleases. Proc. Nat. Acad. Sci. U.S.A. 73: 1102-1106 (1976).

Mintz, B., and Illmensee, K.: Normal genetically mosaic mice produced from malignant teratocarcinoma cells. Proc. Nat. Acad. Sci. U.S.A. 72: 3585-3589 (1975).

Papaioannou, V.E., McBurney, M.W., Gardner, R.L., and Evans, M.J.: Fate of teratocarcinoma cells injected into early mouse embryos. Nature 258: 70-73 (1975).

Pincus, T., Hartley, J.W., and Rowe, W.P.: A major genetic locus affecting resistance to infection with murine leukemia viruses. I. Tissue culture studies of naturally occurring viruses. J. Exp. Med. 133: 1219-1233 (1971).
Sveda, M.M., Fields, B.N., and Soeiro, R.: Host restriction of Friend leukemia virus: fate of input virion RNA. Cell 2: 271-277 (1974).
Tennant, R.W., Schluter, B., Yang, W.-K., and Brown, A.: Reciprocal inhibition of mouse leukemia virus infection by Fv-1 allele cell extracts. Proc. Nat. Acad. Sci. U.S.A. 71: 4241-4245 (1974).
Wiener, F., Klein, G., and Harris, H.: The analysis of malignancy by cell fusion. V. Further evidence of the ability of normal diploid cells to suppress malignancy. J. Cell Sci. 15: 177-183 (1974).
Wiener, F., Klein, G., and Harris, H.: The analysis of malignancy by cell fusion. VI. Hybrids between different tumour cells. J. Cell Sci. 16: 189-198 (1974).

RESEARCH ON CELLULAR SENESCENCE: PRESENT AND FUTURE DIRECTIONS

John W. Littlefield, Wai Nang Choy and Joshua Epstein

Department of Pediatrics, Johns Hopkins University

School of Medicine, Baltimore, Maryland 21205

In this last talk of this workshop, a few points which others have made will be repeated and expanded, and then some aspects of cellular senescence which call for attention in the future will be mentioned briefly.

First, Dr. Sinex made the interesting suggestion that the mechanism of aging may differ for short-lived and long-lived organisms. Perhaps a program for terminal differentiation may occur in the former and a stochastic accumulation of errors in the latter. One wonders, for example, if the pathology of aging is different in mouse and man? Also, could this difference relate to the different behavior of mouse and human fibroblasts in regard to spontaneous transformation? Thus, mouse cells might tolerate mutations because during the life of the mouse not enough mutations could occur to interfere with reproduction, in contrast to human cells. In fact, mouse fibroblasts are much less efficient than human cells in regard to DNA repair after ultraviolet irradiation (1).

It is perplexing that neither spontaneous nor chemically-induced transformation of human fibroblasts has been achieved as yet, despite much effort in many laboratories in recent years. For example, we treated a total of 10^8 fibroblasts from individuals predisposed to tumors, including Down syndrome, Fanconi anemia, and neurofibromatosis, with several agents, including nitrosoguanidine, ethylmethane sulfonate, 5-bromodeoxyuridine, and colchicine. After a variable number of cell generations for expression, we screened for transformed survivors able to grow in soft agar. None were detected. Recently it has been reported that fibroblasts grown from the fibromas of a patient with neurofibromatosis can be transformed by treatment with urethane (2). Success in the chemical transformation of human fibroblasts will probably be achieved soon, but the question will remain why these cells are so much more resistant than the fibroblasts of mice, hamsters, and rats. The different capacity for DNA repair in these cells seems a possible explanation.

Next, Dr. Cristofalo reviewed various data which do not support the Orgel "catastrophe" hypothesis. Also he described his failure to confirm the reported prolongation of the life-span of fibroblasts with vitamin E. For better or worse, we all have some work left to do! We were also reminded that human fibroblasts do not normally seem to need in vivo all the divisions which they can perform in vitro. However, this may not be the case with the several premature aging syndromes, such as Werner syndrome, progeria, the trisomies, and the chromosome breakage syndromes, including Fanconi anemia, Bloom syndrome, and ataxia-telangiectasia, which appear to have a short in vitro lifetime in our experience. In these disorders fibroblasts in vitro may encounter exhaustion of their total division potential. This may be true also for diabetes mellitus, a common disorder which leads to a shortened lifespan in vivo and in vitro (3, 4, 5). Indeed, even certain environmental insults to the fetus during pregnancy, such as rubella and other viruses as well as certain teratogens, might wipe out enough cells (6) so that those remaining were required to divide frequently enough to reach the limit.

Incidentally, it is becoming apparent that the common denominator in many of these inherited premature aging conditions is, once again, defective DNA repair of some type (7).

Several of the speakers and discussants emphasized the extensive heterogeneity present in aging fibroblast cultures. We have come a long way in recent years in recognizing this. First, we interpreted aging cell populations as homogeneous, then we recognized the need to study individual clones, and now we have begun to examine single cells (8, 9). We must go on to record and investigate the subsequent behavior of individual cells in regard to division, as done in the recent studies of Absher et al (10) and Stidworthy (11). These investigators have documented by time lapse cinemicrophotography impressive heterogeneity between daughter cells in regard to the period of time between divisions. Does this relate to different cell size, or different contacts with other cells? Does stem cell behavior ever occur, as can be produced in amebae (12)? When is the non-cycling state temporary, and when permanent, and does this perhaps correlate with a G_1 or G_2 content of DNA? Does SV40 virus perhaps rescue only the temporarily non-cycling cell?

Lately our laboratory has become interested in the production of a temporary non-cycling state in human lymphoblasts growing in suspension (13). These cells have the advantage over fibroblasts that growth can be continuous and logarithmic, without repeated interruptions by confluency and then trypsinization, which lead to artificial bursts of growth. After exposure of lymphoblasts to high concentrations of thymidine, as used for synchrony experiments, 40-70% of the cells appear to enter a temporary non-cycling state (13). Dr. Choy is trying to increase this to 100%, but so far has been unable to do so by prolonged thymidine block, starvation for an amino acid (arginine) either alone or in combination with a thymidine block, or low serum concentration. He is also observing the behavior of individual cells in regard to duration of the non-cycling state and timing of return to cycling behavior.

In recent years the biochemistry of temporarily non-cycling cells, particularly fibroblasts in the confluent state, has received increasing attention, but the pathology of non-cycling cells has barely been touched. At the recent pediatric meetings in St. Louis, McInnes, Shih and Erbe reported an investigation of fibroblasts from patients with Lowe syndrome (oculocerebrorenal syndrome) (14). They noticed that these cells could not survive in the confluent state for much more than a week, in contrast to cells from control patients, which survived over three weeks. Eventually they were able to identify the problem as a defect in glutamate metabolism. Other disorders may be evident in fibroblasts by abnormal behavior only during the non-cycling state. For example, fibroblasts from patients with some of the hereditary degenerative disorders of the muscular and nervous systems might show the effect of the abnormal gene only during the non-cycling state.

At this workshop we have heard described several interesting new techniques. Dr. Norwood mentioned the work at the University of Washington on the visual selection of tetraploid fibroblasts on the basis of the size of the metaphase plate (15). Drs. Yanishevsky and Muggleton-Harris both mentioned the use of gradients to clean up the nuclei which are obtained by cytochalasin enucleation. Dr. Davidson described improvements on the polyethylene glycol fusion technique of Pontecorvo (16), and also a "technique in want of a project", the "superkilling" effect of the combination of 5-bromodeoxyuridine, Hoechst dye, and ultraviolet light. Perhaps this could be used to separate rare non-cycling cells.

Dr. Davidson also mentioned his two remarkable sublines of hamster cells, one of which grows happily despite 100% substitution of thymidine by 5-bromodeoxyuridine and consequent abnormalities in the composition of enzymes and of DNA, and another subline which mysteriously even prefers 5-bromodeoxyuridine to thymidine.

He also made a dramatic generalization about thymidine kinase, which we would rephrase to say that the absence of thymidine kinase activity in a cell extract often correlates with resistance to 5-bromodeoxyuridine and sensitivity to HAT medium, but does not always do so. As Dr. Davidson pointed out, cells may be able to survive in HAT and/or sensitive to 5-bromodeoxyuridine even when enzyme activity is absent in extracts, and vice versa.

Concerning cell fusion, three of the speakers and discussants described attempts to fuse cytoplasm with senescent fibroblasts in order to determine if rescue occurs. The conclusions of this work, and that elsewhere (17), seem uncertain as yet. In regard to heterokaryons, it is surprising to learn from Dr. Norwood that after 2-3 days HeLa cells lose their ability to activate DNA synthesis in a senescent nucleus, in contrast to cells transformed with SV40 virus. What is more powerful about the SV40 transformed cells in this regard? Furthermore, Dr. Yanishevsky found that an established line of cells, which were better able to enter the non-cycling state than HeLa cells or SV40 transformed cells, did not stimulate DNA synthesis in senescent nuclei at all! Surely she will explore this distinction further, trying other cell lines which are both aneuploid and able to enter a non-cycling state, such as 3T3.

Concerning viable hybrid cells, our own work (18) and recent preliminary studies by the Seattle group suggest that senescence is dominant in crosses with young fibroblasts, in contrast to published data on crosses with established lines. At least in regard to cells transformed with SV40 virus, Dr. Croce has shown that rescue may require the continued presence of the viral genome. This seems to reside on chromosome no. 7 in the case of three independently transformed lines, but on chromosome no. 17 in a fourth line. We need to construct other hybrids with other established human lines, including those which have transformed spontaneously or after treatment with carcinogens, to see if different chromosomes can be responsible for the rescue of senescent cells. Also it remains possible that in some cases one or more chromosomes containing "senescent" loci may have been lost from the fibroblast part of the hybrid genome, and specific hybrids should be constructed to examine this possibility.

Some topics seem to have received less emphasis at this workshop, in contrast with past, and very likely, future meetings. Thus, we have heard little about the possible study of senescence in other cell types, except for comments by Drs. Nichols and Miller in regard to ploidy changes in epithelial cells. These cells may be especially useful for studies on senescence, because they are particularly short-lived. Techniques are becoming available for the better handling of epithelial cells in culture, derived from skin, amnion, kidney, anmiotic fluid, etc. (19, 20). For example, Rheinwald and Green recently published a feeder layer technique which extends the lifetime and promotes the keratinization of skin epithelial cells (21). They obtained the impression that there was an inverse proportion between the lifespan of these cells and the age of the donor, as in fibroblasts, but more extensive studies need to be done. Too, a few studies have appeared on chemical carcinogenesis in epithelial cell cultures (22, 23).

It is fascinating that all diploid cells appear to run out of divisions sooner or later, albeit at different times. This includes epithelial cells and lymphocytes, as well as fibroblasts. When any of these cells are stimulated by viruses, such as SV40 virus or Epstein-Barr virus, they go through more divisions, but eventually they again enter a senescent stage or "crisis", with frequent chromosome abnormalities. This is followed by the recovery of a few cells to establish a heteroploid line. The scenario is the same but the tempo is different for the different cell types (24).

Indeed, more discussion of the role of chromosome abnormalities in senescence seems desirable. This also was mentioned briefly and only by Drs. Nichols and Miller. Chromosome abnormalities in cellular senescence received considerable attention ten years ago. The possibility remains that they may be much more important in producing senescence than we realize. Many may not be scored because they prevent re-entry of the cell into mitosis and therefore are never recognized. Monosomies may have this effect, for example. Unfortunately Dr. Yanishevsky and Dr. Schneider did not discuss the current status of their interesting studies on this subject (8, 9). Each of these investigators has shown that there is increased variation in the DNA content of non-dividing cells in older cultures. We and others (25) have considered using the technique of premature chromosome condensation (26) to examine the chromosome complement of the non-dividing cells in a senescent culture. Dr. Epstein has now set up this technique (Fig. 1). We find that the problem is getting enough well-spread metaphases to do accurate counts and karyotyping.

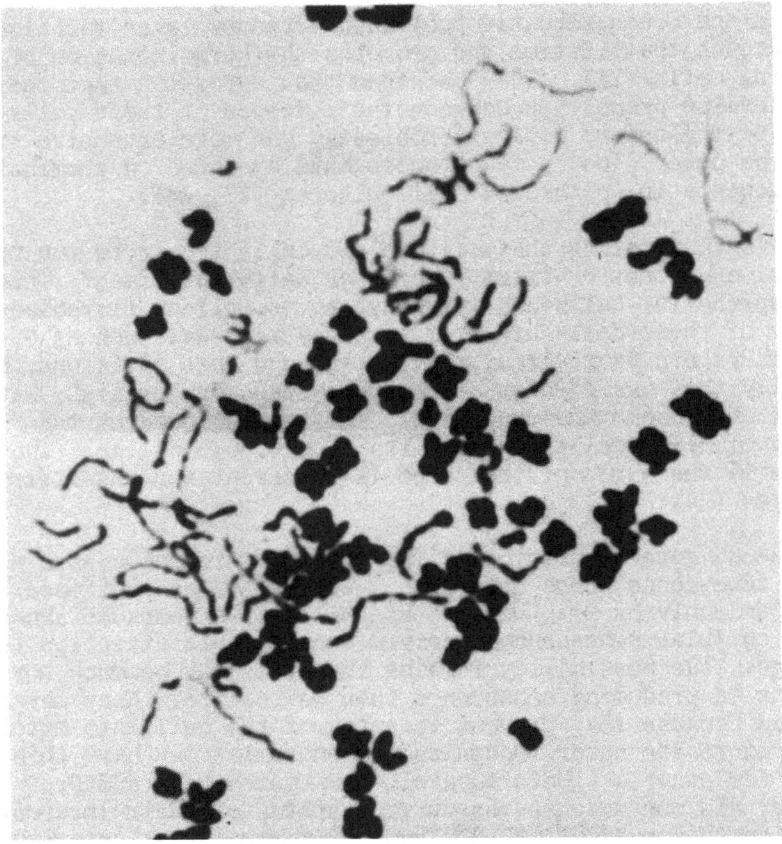

Fig. 1. Premature chromosome condensation. HeLa cells, growing at a concentration of 2 x 10^5/ml in spinner medium containing 5% fetal calf serum, were blocked with thymidine (2 x 10^{-4} M), and transferred to petri dishes at 37°. The thymidine block was released at 16 hours. At 20 hours a nitrous oxide block (80 psi) was applied, and then released at 29 hours. At this time the mitotic index of the cells floating in the medium was 90% or better. Then 10^7 metaphase HeLa cells and 5 x 10^6 human fibroblasts, collected from confluent dishes by trypsinization and resuspended in medium, were mixed and centrifuged. The combined cells were washed once in medium lacking fetal calf serum but containing Colcemid (2 micrograms/ml). Sendai virus was added to the pellet for 15 minutes on ice and 45 minutes at 37°, with very gentle shaking. Then 4 volumes of water at room temperature was added for 20 minutes, and chromosome preparations prepared as usual. In the figure are short dark HeLa chromosomes mixed with prematurely condensed single chromatids from a fibroblast.

The potential role of chromosome abnormalities in cancer also is of great interest. Cellular senescence and cancer are closely related, since cancer always involves the overcoming of senescence. As mentioned previously, in mouse fibroblast cultures spontaneous transformation regularly occurs to establish a permanent cell line. Transformation seems equivalent to the development of neoplasia, or at least a major first step in this direction. Review of the important studies of Todaro and Green (27) and of the Toronto group (28) suggests that the time of establishment of mouse cell cultures coincides closely with the acquisition of heteroploidy. Four relations between heteroploidy and neoplasia seem possible: "1. Heteroploidy is a secondary acquisition after establishment, selected for because it promotes variation and facilitates progression by offering many new phenotypes; 2. Heteroploidy accompanies establishment simply because established cells have passed through and escaped senescence, in which chromosome abnormalities are a common feature; 3. Heteroploidy is a concomitant pleiotropic effect of the event which causes establishment; 4. Heteroploidy is the event causing establishment" (24).

The last relation is the simplest. Indeed a universal scenario can be proposed, which is not especially new, but which certainly is in need of careful consideration: "Senescence may appear eventually in all diploid cells due to programmed or stochastic events causing abnormal chromosome replication or segregation. Rarely, such changes would be compatible with continued growth and division; a cell that could continue to grow would be under strong selective pressure, and would be regarded as 'established,' that is, as having escaped senescence. the cell would also continue to generate new chromosome rearrangements, and therefore would be 'heteroploid'. Sooner or later, some such rearrangement could well cause diminished growth control" (24).

Acknowledgment

Research studies mentioned in this paper were supported by U.S.P.H.S. research grant no. 5 R01 CA 16754. In addition Dr. Choy has been supported by U.S.P.H.S. training grant no. 5 T22 GM 00145 and Dr. Epstein by U.S.P.H.S. postdoctoral fellowship no. SF22 Ag0503602.

References

1. Hart, R.W., and Setlow, R.B. Correlation between deoxyribonucleic acid excision-repair and life-span in a number of mammalian species. Proc. Natl. Acad. Sci. U.S.A., *71*:2169-73, 1974.

2. Benedict, W.F., Jones, P.A., Laug, W.E., Igel, H.J., and Freeman, A.E. Characterization of human cells transformed *in vitro* by urethane. Nature, *256*:322-324, 1975.

3. Goldstein, S., Littlefield, J.W., and Soeldner, J.S. Diabetes mellitus and aging: Diminished plating efficiency of cultured human fibroblasts. Proc. Natl. Acad. Sci. U.S.A., *64*:155-160, 1969.

4. Goldstein, S., in preparation.

5. Vracko, R., and Benditt, E.P. Restricted replicative life-span of diabetic fibroblasts *in vitro*: its relation to microangiopathy. Fed. Proc., *341*:68-70, 1975.

6. Naeye, R.L., and Blanc, W. Pathogenesis of congenital rubella. J.A.M.A., *194*:1277-83, 1965.

7. Brown, W.T., Epstein, J., and Little, J.B. Progeria cells are stimulated to repair DNA by co-cultivation with normal cells. Exper. Cell Res., *97*:291-296, 1976.

8. Yanishevsky, R., Mendelsohn, M.L., Mayall, B.H., and Cristofalo, V.J. Proliferative capacity and DNA content of aging human diploid cells in culture: a cytophotometric and autoradiographic analysis. J. Cell. Pbysiol., *84*:165-170, 1974.

9. Schneider, E.L, and Fowlkes, B.J. Measurement of DNA content and cell volume in senescent human fibroblasts utilizing flow multiparameter single cell analysis. Exper. Cell Res., *98*:298-302, 1976.

10. Absher, P.M., Absher, R.G., and Barnes, W.D. Genealogies of clones of diploid fibroblasts. Exper. Cell Res., *88*:94-104, 1974.

11. Stidworthy, G.H. Cinemicrophotographic analysis of cell division of normal human fibroblasts during *in vitro* aging. In Vitro, *12*:301, 1976.

12. Danielli, J.F., and Muggleton, A. Some alternative states of amoeba, with special reference to life-span. Gerontologia,

3:76-90, 1959.

13. Zielke, H.R., and Littlefield, J.W. Repetitive synchronization of human lymphoblast cultures with excess thymidine. D.M. Prescott, ed., Methods in Cell Biology, Vol. 8, pp. 107-121. New York, Academic Press.

14. McInnes, R.R., Shih, V.E., and Erbe, R.W. Metabolic abnormalities in Lowe syndrome fibroblasts. Ped. Res., 10:368, 1976.

15. Hoehn, H., Bryant, E.M., Johnston, P., Norwood, T.H., and Martin, G.M. Non-selective isolation, stability and longevity of hybrids between normal human somatic cells. Nature, 258:608-610, 1975.

16. Davidson, R.L. and Gerald, P.S. Improved techniques for the induction of mammalian cell hybridization by polyethylene glycol. Somatic Cell Genetics, 2:165-176, 1976.

17. Wright, W.E., and Hayflick, L. Nuclear control of cellular aging demonstrated by hybridization of anucleate and whole cultured normal human fibroblasts. Exper. Cell Res., 96:113-121, 1975.

18. Littlefield, J.W. Attempted hybridizations with senescent human fibroblasts. J. Cell Physiol., 82:129-131, 1973.

19. Flaxman, B.A. Cell identification in primary cell cultures from skin. In Vitro, 10:112-118, 1974.

20. Gilbert, S.F., and Migeon, B.R. D-valine as a selective agent for normal human and rodent epithelial cells in culture. Cell, 5:11-17, 1975.

21. Rheinwald, J.G., and Green, H. Serial cultivation of strains of human epidermal keratinocytes: the formation of keratinizing colonies from single cells. Cell, 6:331-344, 1975.

22. Fusenig, N.E., Samsel, W., Thon, W., and Worst, P.K.M. Malignant transformation of epidermal cells in culture by DMBA. INSERM, 19:219-228, 1973.

23. Hashimoto, Y., and Kitagawa, H.S. In vitro neoplastic transformation of epithelial cells of rat urinary bladder by nitrosamines. Nature, 252:497-499, 1974.

24. Littlefield, J.W. "Variation, Senescence and Neoplasia in Cultured Somatic Cells". Harvard University Press, Cambridge,

Mass., 1976.

25. Yanishevsky, R., and Carrano, A.V. Prematurely condensed chromosomes of dividing and non-dividing cells in aging human cell cultures. Exper. Cell Res., 90:169-174, 1975.

26. Rao, P.N., and Johnson, R.T. Mammalian cell fusion: studies on the regulation of DNA synthesis and mitosis. Nature, 225:159-164, 1970.

27. Todaro, G.J., and Green, H. Quantitative studies of the growth of mouse embryo cells in culture and their development into established lines. J. Cell Biol., 17:299-313, 1963.

28. Rothfels, K.H., Kupelwieser, E.B., and Parker, R.C. Effects of X-irradiated feeder layers on mitotic activity and development of aneuploidy in mouse-embryo cells in vitro. Canadian Cancer Conference, 5:191-223, 1963.

SUBJECT INDEX

MIX
Papier aus verantwortungsvollen Quellen
Paper from responsible sources
FSC® C105338

If you have any concerns about our products,
you can contact us on
ProductSafety@springernature.com

In case Publisher is established outside the EU,
the EU authorized representative is:
Springer Nature Customer Service Center GmbH
Europaplatz 3, 69115 Heidelberg, Germany

Printed by Libri Plureos GmbH
in Hamburg, Germany